Lecture Notes in Computer S

Edited by G. Goos, J. Hartmanis and J. van Leeuwen

Springer

Berlin
Heidelberg
New York
Barcelona
Hong Kong
London
Milan
Paris
Singapore
Tokyo

Gerth Stølting Brodal   Daniele Frigioni
Alberto Marchetti-Spaccamela (Eds.)

Algorithm Engineering

5th International Workshop, WAE 2001
Århus, Denmark, August 28-31, 2001
Proceedings

 Springer

Series Editors

Gerhard Goos, Karlsruhe University, Germany
Juris Hartmanis, Cornell University, NY, USA
Jan van Leeuwen, Utrecht University, The Netherlands

Volume Editors

Gerth Stølting Brodal
University of Aarhus, BRICS, Department of Computer Science
8000 Åarhus, Denmark
E-mail: gerth@brics.dk

Daniele Frigioni
Universitá dell'Aquila, Dipartimento di Ingegneria Elettrica
Poggio di Roio, 67040 L'Aquila, Italy
E-mail: frigioni@ing.univaq.it

Alberto Marchetti-Spaccamela
Universitá di Roma "La Sapienza", Dipartimento di Informatica e Sistemistica
via Salaria 113, 00198 Roma, Italy
E-mail: alberto@dis.uniroma1.it

Cataloging-in-Publication Data applied for

Die Deutsche Bibliothek - CIP-Einheitsaufnahme

Algorithm engineering : 5th international workshop ; proceedings / WAE 2001,
Aarhus, Denmark, August 28 - 31, 2001. Gerth Stølting Brodal ... (ed.). -
Berlin ; Heidelberg ; New York ; Barcelona ; Hong Kong ; London ; Milan ;
Paris ; Singapore ; Tokyo : Springer, 2001
   (Lecture notes in computer science ; Vol. 2141)
   ISBN 3-540-42500-4

CR Subject Classification (1998): F.2, G.2, E.1, C.2, G.1

ISSN 0302-9743
ISBN 3-540-42500-4 Springer-Verlag Berlin Heidelberg New York

Springer-Verlag Berlin Heidelberg New York
a member of BertelsmannSpringer Science+Business Media GmbH

http://www.springer.de

© Springer-Verlag Berlin Heidelberg 2001
Printed in Germany

Typesetting: Camera-ready by author, DA-TeX Gerd Blumenstein
Printed on acid-free paper   SPIN 10840177   06/3142   5 4 3 2 1 0

Preface

This volume contains the papers accepted for presentation at the *5th Workshop on Algorithm Engineering* (WAE 2001) held in Århus, Denmark, on August 28–31, 2001. The Workshop on Algorithm Engineering is an annual meeting devoted to researchers and developers interested in practical aspects of algorithms and their implementation issues. The goal of the workshop is to present recent research results and to identify and explore directions of future research in the field of algorithm engineering. Previous meetings were held in Venice (1997), Saarbrücken (1998), London (1999), and Saarbrücken (2000).

Papers were solicited describing original research in all aspects of algorithm engineering including:

- Implementation, experimental testing, and fine-tuning of discrete algorithms.
- Development of software repositories and platforms which allow use of, and experimentation with, efficient discrete algorithms.
- Novel uses of discrete algorithms in other disciplines and the evaluation of algorithms for realistic environments.
- Methodological issues including standards in the context of empirical research on algorithms and data structures.
- Methodological issues regarding the process of converting user requirements into efficient algorithmic solutions and implementations.

The Program Committee selected 15 papers from a total of 25 submissions from 17 countries, according to selection criteria, taking into account paper originality, quality, and relevance to the workshop.

WAE 2001 was jointly organized with ESA 2001 (the *9th European Symposium on Algorithms*) and WABI 2001 (the *1st Workshop on Algorithm in BioInformatics*) in the context of ALGO 2001. ALGO 2001 had seven invited talks most of which were relevant to WAE 2001. The seven distinguished invited speakers of ALGO 2001 were:

Susanne Albers	University of Dortmund
Lars Arge	Duke University
Andrei Broder	AltaVista
Herbert Edelsbrunner	Duke University
Jotun Hein	University of Aarhus
Gene Myers	Celera Genomics
Uri Zwick	Tel Aviv University

We want to thank all the people who contributed to the success of the workshop: those who submitted papers for consideration; the program committee members and the referees for their timely and invaluable contribution; the invited speakers of ALGO 2001; the members of the organizing committee for their dedicated work; BRICS (the center for Basic Research In Computer Science, University of Aarhus) and EATCS (the European Association for Theoretical Computer Science) for financial support. We also thank ACM SIGACT (the Association for Computing Machinery Special Interest Group on Algorithms and Computation Theory) for providing us the software used for handling the electronic submissions and the electronic program committee meeting.

June 2001   Gerth Stølting Brodal
                                                       Daniele Frigioni
                                          Alberto Marchetti-Spaccamela

Program Committee

Ulrik Brandes	University of Konstanz
Paolo Ferragina	University of Pisa
Klara Kedem	Ben-Gurion University
Alberto Marchetti-Spaccamela (Chair)	University of Rome "La Sapienza"
Bernard M. E. Moret	University of New Mexico
Mark Overmars	Utrecht University
Stefan Schirra	Think & Solve, Saarbrücken
Monique Teillaud	INRIA - Sophia Antipolis

Organizing Committee

Gerth Stølting Brodal	BRICS - University of Aarhus
Rolf Fagerberg	BRICS - University of Aarhus
Daniele Frigioni	University of L'Aquila
Karen Kjær Møller	BRICS - University of Aarhus
Erik Meineche Schmidt (Chair)	BRICS - University of Aarhus

Additional Referees

Eitan Bachmat	Gabriele Di Stefano	Avraham Melkman
Amos Beimel	Ioannis Emiris	Linda Pagli
Serafino Cicerone	Antonio Frangioni	Marco Pellegrini
Valentina Ciriani	Daniele Frigioni	Andrea Pietracaprina
Fabrizio D'Amore	Roberto Grossi	Sylvain Pion
Frank Dehne	Matthew Katz	Nadia Pisanti
Camil Demetrescu	Jörg Keller	Eyal Shimony
Yefim Dinitz	Giovanni Manzini	

Table of Contents

Compact DFA Representation for Fast Regular Expression Search[*]

Gonzalo Navarro[1] and Mathieu Raffinot[2]

[1] Dept. of Computer Science, University of Chile
Blanco Encalada 2120, Santiago, Chile
gnavarro@dcc.uchile.cl
[2] Equipe génome, cellule et informatique, Université de Versailles
45 avenue des Etats-Unis, 78035 Versailles Cedex
raffinot@genetique.uvsq.fr.

Abstract. We present a new technique to encode a deterministic finite automaton (DFA). Based on the specific properties of Glushkov's nondeterministic finite automaton (NFA) construction algorithm, we are able to encode the DFA using $(m+1)(2^{m+1} + |\Sigma|)$ bits, where m is the number of characters (excluding operator symbols) in the regular expression and Σ is the alphabet. This compares favorably against the worst case of $(m+1)2^{m+1}|\Sigma|$ bits needed by a classical DFA representation and $m(2^{2m+1} + |\Sigma|)$ bits needed by the Wu and Manber approach implemented in *Agrep*.

Our approach is practical and simple to implement, and it permits searching regular expressions of moderate size (which include most cases of interest) faster than with any previously existing algorithm, as we show experimentally.

1 Introduction and Related Work

The need to search for regular expressions arises in many text-based applications, such as text retrieval, text editing and computational biology, to name a few. A *regular expression* is a generalized pattern composed of (i) basic strings, (ii) union, concatenation and Kleene closure of other regular expressions. Readers unfamiliar with the concept and terminology related to regular expressions are referred to a classical book such as [ ]. We call RE our regular expression, which is of length m. This means that m is the total number of characters in RE, not counting operators symbols "|", "*" and parentheses. We note $L(RE)$ the set of words generated by RE and Σ the alphabet.

The traditional technique [ ] to search a regular expression of length m in a text of length n is to convert the expression into a nondeterministic finite automaton (NFA) with $O(m)$ nodes. Then, it is possible to search the text using the automaton at $O(mn)$ worst case time. The cost comes from the fact that

[*] Partially supported by ECOS-Sud project C99E04 and, for the first author, Fondecyt grant 1-990627.

G. Brodal et al. (Eds.): WAE 2001, LNCS 2141, pp. 1–13, 2001.

more than one state of the NFA may be active at each step, and therefore all may need to be updated. A more efficient choice [ ] is to convert the NFA into a deterministic finite automaton (DFA), which has only one active state at a time and therefore allows searching the text at $O(n)$ cost, which is worst-case optimal. The problem with this approach is that the DFA may have $O(2^m)$ states, which implies a preprocessing cost and extra space exponential in m.

Some techniques have been proposed to obtain a good tradeoff between both extremes. In 1992, Myers [ ] presented a four-russians approach which obtains $O(mn/\log n)$ worst-case time and extra space. The idea is to divide the syntax tree of the regular expression into "modules", which are subtrees of a reasonable size. These subtrees are implemented as DFAs and are thereafter considered as leaf nodes in the syntax tree. The process continues with this reduced tree until a single final module is obtained.

The DFA simulation of modules is done using *bit-parallelism*, which is a technique to code many elements in the bits of a single computer word (which is called a "bit mask") and manage to update all them in a single operation. Typical bit operations are infix "|" (bitwise *or*), infix "&" (bitwise *and*), prefix "~" (bit complementation), and infix "<<" (">>"), which moves the bits of the first argument (a bit mask) to higher (lower) positions in an amount given by the right argument. Additionally, one can treat the bit masks as numbers and obtain specific effects using the arithmetic operations $+$, $-$, etc. Exponentiation is used to denote bit repetition, e.g. $0^3 1 = 0001$.

In our case, the vector of active and inactive states is stored as bits of a computer word. Instead of (ala Thompson [ ]) examining the active states one by one, the whole computer word is used to index a table which, together with the current text character, provides the new bit mask of active states. This can be considered either as a bit-parallel simulation of an NFA, or as an implementation of a DFA (where the identifier of each deterministic state is the bit mask as a whole).

Pushing even more on this direction, one may resort to pure bit-parallelism and forget about the modules. This was done in [ ] by Wu and Manber, and included in their software *Agrep* [ ]. A computer word is used to represent the active (1) and inactive (0) states of the NFA. If the states are properly arranged and the Thompson construction [ ] is used, then all the arrows carry 1's from bit positions i to $i + 1$, except for the ε-transitions. Then, a generalization of Shift-Or [ ] (the canonical bit-parallel algorithm for exact string matching) is presented, where for each text character two steps are performed. First, a forward step moves all the 1's that can move from a state to the next one. This is achieved by precomputing a table $B : \Sigma \to 2^{O(m)}$, such that the i-th bit of $B[c]$ is set if and only if the character c matches at the i-th position of the regular expression. Second, the ε-transitions are carried out. As ε-transitions follow arbitrary paths, a table $E : 2^{O(m)} \to 2^{O(m)}$ is precomputed, where $E[D]$ is the ε-closure of D. To move from the state set D to the new D' after reading text character c, the action is

$$D' \;\leftarrow\; E[(D << 1) \mid B[c]].$$

Possible space problems are solved by splitting this table "horizontally" (i.e. less bits per entry) in as many subtables as needed, using the fact that $E[D_1 D_2] = E[D_1 0^{|D_2|}] \mid E[0^{|D_1|} D_2]$. This can be thought of as an alternative decomposition scheme, instead of Myers' modules.

All the approaches mentioned are based on the Thompson construction of the NFA, whose properties have been exploited in different ways. An alternative, much less known, NFA construction algorithm is Glushkov's [,]. A good point of this construction is that, for a regular expression of m characters, the NFA obtained has exactly $m + 1$ states and is free of ε-transitions. Thompson's construction, instead, produces between $m + 1$ and $2m$ states. This means that Wu and Manber's table may need a table of size 2^{2m} entries of $2m$ bits each, for a total space requirement of $m(2^{2m+1} + |\Sigma|)$ bits (E plus B tables).

In [ ], we proposed the use of Glushkov's construction instead of Thompson's. The table had then 2^{m+1} entries, but unfortunately the structural property that arrows were either forward or ε-transitions did not hold anymore. As a result, we needed a table $M : 2^{m+1} \times \Sigma \to 2^{m+1}$ indexed by the current state and text character, for a total space requirement of $(m + 1)2^{m+1}|\Sigma|$ bits. The transition action was simply $D' \leftarrow M[D, c]$, just as for a classical DFA implementation. We showed experimentally that the Glushkov based construction was normally faster than the one based on Thompson, but not better than a classical DFA.

In this paper, we use specific properties of the Glushkov construction (namely, that all the arrows arriving to a state are labeled by the same letter) to eliminate the need of a separate table per text character. As a result, we obtain the best of both worlds: we can have tables whose arguments have just $m+1$ bits and we can have just one table instead of one per character. Thus we can represent the DFA using $(m + 1)(2^{m+1} + |\Sigma|)$ bits, which is not only better than both previous bit parallel implementations but also better than the classical DFA representation, which needs in the worst case $(m + 1)2^{m+1}|\Sigma|$ bits.

The net result is a simple algorithm for regular expression searching which uses normally less space and has faster preprocessing and search time (albeit all are $O(n)$ search time, a smaller DFA representation implies more locality of reference). We show experimentally that we are at least 10% faster than any previous algorithm for searching regular expressions of moderate size, which include most cases of interest.

The algorithms reviewed are called "forward scanning" algorithms because they inspect all the text characters, one by one. It should be noted that there exist algorithms able to skip text characters, which discard text areas that cannot contain a match and use a classical algorithm on the rest, e.g. [,] and that of *Gnu Grep v2.0*. Those algorithms are in some cases faster than ours, but all them need a forward scan algorithm to search the text areas that cannot be discarded. Hence, a better forward scanning algorithm is always welcome. Moreover, many interesting regular expressions cannot be efficiently searched using backward scanning algorithms. We consider in the Conclusions how to use our technique directly in some of those algorithms too.

2   Glushkov Automaton

There exist currently many different techniques to build an NFA from a regular expression RE of m characters (without counting the special symbols). The most classical one is the Thompson construction [ ], which builds an NFA with at most $2m$ states (and at least $m + 1$). This NFA has some particular properties (e.g. $O(1)$ transitions leaving each node) that have been extensively exploited in several regular expression search algorithm such as that of Thompson [ ], Myers [ ] and Wu and Manber [,].

Another particularly interesting NFA construction algorithm is by Glushkov [ ], popularized by Berry and Sethi in [ ]. The NFA resulting from this construction has the advantage of having just $m + 1$ states (one per position in the regular expression). Its number of transitions is worst case quadratic, but this is unimportant under our bit-parallel representation (it just means denser bit masks). We present this construction in depth.

2.1   Glushkov Construction

The construction begins by marking the positions of the characters of Σ in RE, counting only characters. For instance, $(\texttt{AT}|\texttt{GA})((\texttt{AG}|\texttt{AAA})*)$ is marked $(A_1 T_2 | G_3 A_4)((A_5 G_6 | A_7 A_8 A_9)*)$. A *marked expression* from a regular expression RE is denoted $\overline{RE}$ and its language (including the indices on each character) $L(\overline{RE})$. On our example, $L((A_1 T_2 | G_3 A_4)((A_5 G_6 | A_7 A_8 A_9)*)) = \{A_1 T_2,\ G_3 A_4,\ A_1 T_2 A_5 G_6,\ G_3 A_4 A_5 G_6,\ A_1 T_2 A_7 A_8 A_9,\ G_3 A_4 A_7 A_8 A_9,\ A_1 T_2 A_5 G_6 A_5 G_6, \ldots\}$. Let $Pos(\overline{RE})$ be the set of positions in $\overline{RE}$ (*i.e.*, $Pos = \{1 \ldots m\}$) and $\overline{\Sigma}$ the marked character alphabet.

The Glushkov automaton is built first on the marked expression $\overline{RE}$ and it recognizes $L(\overline{RE})$. We then derive from it the Glushkov automaton that recognizes $L(RE)$ by erasing the position indices of all the characters (see below).

The idea of Glushkov is the following. The set of positions is taken as a reference, becoming the set of states of the resulting automaton (adding an initial state 0). So we build $m + 1$ states labeled from 0 to m. Each state j represents the fact that we have read in the text a string that ends at NFA position j. Now if we read a new character σ, we need to know which positions $\{j_1 \ldots j_k\}$ we can reach from j by σ. Glushkov computes from a position (state) j all the other accessible positions $\{j_1 \ldots j_k\}$.

We need four new definitions to explain in depth the algorithm. We denote below by σ_y the indexed character of $\overline{RE}$ that is at position y.

Definition $First(\overline{RE}) = \{x \in Pos(\overline{RE}),\ \exists u \in \overline{\Sigma}^*,\ \sigma_x u \in L(\overline{RE})\}$, *i.e. the set of initial positions of $L(\overline{RE})$, that is, the set of positions at which the reading can start. In our example, $First((A_1 T_2 | G_3 A_4)((A_5 G_6 | A_7 A_8 A_9)*)) = \{1, 3\})$.*

Definition $Last(\overline{RE}) = \{x \in Pos(\overline{RE}),\ \exists u \in \overline{\Sigma}^*,\ u \sigma_x \in L(\overline{RE})\}$, *i.e. the set of final positions of $L(\overline{RE})$, that is, the set of positions at which a string read* can be recognized. In our example, $Last((A_1T_2|G_3A_4)((A_5G_6|A_7A_8A_9)*)) = \{2, 4, 6, 9\})$.

Definition $Follow(x) = \{y \in Pos(\overline{RE}), \exists u, v \in \Sigma^*, u\sigma_x\sigma_y v \in L(\overline{RE})\}$, i.e. all the positions in $Pos(\overline{RE})$ accessible from x. For instance, in our example, if we consider position 6, the set of accessible positions $Follow((A_1T_2|G_3A_4)((A_5G_6| A_7A_8A_9)*), 6) = \{7, 5\}$.

Definition $Empty_{RE}$ is $\{\varepsilon\}$ if ε belongs to $L(RE)$ and $\emptyset$ otherwise.

The Glushkov automaton $\overline{GL} = (S, \Sigma, I, F, \overline{\delta})$ that recognizes the language $L(\overline{RE})$ is built from these three sets in the following way (Figure 1 shows our example NFA).

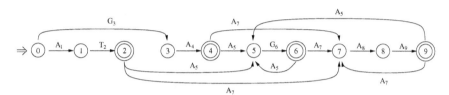

Fig. 1. Marked Glushkov automaton built on the marked regular expression $(A_1T_2|G_3A_4)((A_5G_6|A_7A_8A_9)*)$. The state 0 is initial. Double-circled states are final

1. S is the set of states, $S = \{0, 1, \ldots, m\}$, i.e., the set of positions $Pos(\overline{RE})$ and the initial state is $I = 0$.
2. F is the set of final states, $F = Last(\overline{RE}) \cup (Empty_{RE} \cdot \{0\})$. Informally, a state (position) i is final if it is in $Last(\overline{RE})$ (in which case when reaching such a position we know that we recognized a string in $L(\overline{RE})$). The initial state 0 is also final if the empty word ε belongs to $L(\overline{RE})$, in which case $Empty_{RE} = \{\varepsilon\}$ and hence $Empty_{RE} \cdot \{0\} = \{0\}$. If not, $Empty_{RE} = Empty_{RE} \cdot \{0\} = \emptyset$.
3. $\overline{\delta}$ is the transition function of the automaton, defined by

$$\forall x \in Pos(\overline{RE}), \; \forall y \in Follow(\overline{RE}, x), \; \overline{\delta}(x, \sigma_y) = y. \tag{1}$$

Informally, there is a transition from state x to y by σ_y if y follows x.

The Glushkov automaton of the original RE is now simply obtained by erasing the position indices in the marked automaton. The new automaton recognizes the language $L(RE)$.

The complexity of this construction is $O(m^3)$, which can be reduced to $O(m^2)$ in different ways by using distinct properties of the $First$ and $Follow$ sets [,]. However, when using bit parallelism, the complexity is directly reduced to $O(m^2)$ by manipulating all the states in a register (see Section 3).

3   DFA Representation and Search Algorithm

The classical algorithm to produce a DFA from an NFA consists in making each DFA state represent a set of NFA states which may be active at that point. A possible way to represent the states of a DFA (i.e. the sets of states of an NFA) is to use a bit mask of $O(m)$ bits, as already explained. Previous bit-parallel implementations [?, ?] are built on this idea. We present in this section a new bit-parallel DFA representation based on Glushkov's construction. As we make heavy use of this construction and its properties, we start by presenting a bit-parallel implementation of Glushkov's construction.

3.1   Bit-Parallel Glushkov Construction

All along the Glushkov algorithm we manipulate sets of NFA states. As it is useful for the search algorithm, we will use bit-parallelism to represent these sets of states, that is, we will represent sets using bit masks of $m + 1$ bits, where the i-th bit is 1 if and only if state number i belongs to the set.

An immediate advantage of using a bit-parallel implementation is that we can easily handle *classes of characters*. This means that at each position of the regular expression there is not just a character of Σ but a set of characters, any of which is good to traverse the corresponding arrow. Rather than just converting the set $\{a_1, a_2, \ldots a_k\}$ into $(a_1 | a_2 | \ldots | a_k)$ (and creating k positions instead of one), we can consider the class as a single letter.

The algorithm of Glushkov is based on the parse tree of the regular expression. Each node v of this tree represents a sub-expression RE_v of RE. For each node, its variables $First(v)$, $Last(v)$, $Follow(v, x)$ and $Empty_v$ are computed in postfix order. We will consider that regular expressions contain classes of characters rather than single characters at the leaves of their syntax trees.

Together with the above mentioned variables, we fill a table of bitmasks $B : \Sigma \rightarrow 2^{m+1}$, such that the i-th bit of $B[c]$ is set if and only if c belongs to the class at the i-th position of the regular expression.

3.2   Properties of Glushkov's Construction

We present now some properties of the Glushkov construction which are necessary for our compact DFA representation. All them are very easy to prove.

A first property should be obvious at this point, but it is important because it makes our problem totally different from that of a Thompson's construction: since we do not build any ε-transitions, we have that Glushkov's NFA is ε-free.

That is, in the approach of Wu and Manber [?], the ε-transitions are the complicated part, because all the others move forward. We do not have these transitions in the Glushkov automaton, but on the other hand the normal transitions do not follow such a simple pattern.

However, there are still important structural properties in the arrows. One of these is captured in the following Lemma.

Lemma 1 *All the arrows leading to a given state in Glushkov's NFA are labeled by the same character. Moreover, if classes of characters are permitted at the positions of the regular expression, then all the arrows leading to a given state in Glushkov's NFA are labeled by the same class.*

Proof. This is easily seen in Formula (1). The character labeling the arrows that arrive at state y is precisely σ_y. This also holds if we consider that σ_y is in fact a subset of Σ.

These properties can be combined with the B table to yield our most important property.

Lemma 2 *Let $B(\sigma)$ be the set of positions of the regular expression that contain character σ. Let $Follow(x)$ be the set of states that can follow state x in one transition, by Glushkov's construction. Let $\delta : S \times \Sigma \rightarrow S$ the transition function of the Glushkov's NFA, i.e. $y \in \delta(x, \sigma)$ if and only if from state x we can move to state y by character σ. Then, it holds*

$$\delta(x, \sigma) \quad = \quad Follow(x) \ \cap \ B(\sigma)$$

Proof. The lemma follows from Lemma 1. Let $y \in \delta(x, \sigma)$. This means that y can be reached from x by σ and therefore $y \in Follow(x) \cap B(\sigma)$. Conversely, let $y \in Follow(x) \cap B(\sigma)$. Then y can be reached by letter σ and it can be reached from x. But Lemma 1 implies that every arrow leading to y is labeled by σ, including the one departing from x, and hence $y \in \delta(x, \sigma)$.

Finally, a last property is necessary for technical reasons made clear shortly.

Lemma 3 *The initial state 0 in Glushkov's NFA does not receive any arrow.*

Proof. This is clear since all the arrows are built in Formula (1), and the initial state is not in the $Follow$ set of any other state (see the definition of $Follow$).

3.3   A Compact DFA Representation

We now use Lemma 2 to obtain a compact representation of the DFA. The idea is to compute the transitions by using two tables: the first one is simply $B[\sigma]$, which is built in algorithm **Glushkov_variables** and gives a bit mask of the states reachable by each letter (no matter from where). The second is a deterministic version of $Follow$, i.e. a table T from sets of states to sets of states (in bit mask form) which tells which states can be reached from an active state in D, no matter by which character:

$$T[D] \quad = \quad \bigcup_{i \in D} Follow(i)$$

By Lemma 2, it holds that

$$\delta(D, \sigma) \quad = \quad T[D] \ \& \ B[\sigma]$$

(where we are using bit mask representation for sets). Hence instead of the complete transition table $\delta : 2^{m+1} \times \Sigma \rightarrow 2^{m+1}$ we build and store only $T : 2^{m+1} \rightarrow 2^{m+1}$ and $B : \Sigma \rightarrow 2^{m+1}$. The number of bits required in this representation is $(m+1)(2^{m+1} + |\Sigma|)$. Figure 2 shows the algorithm to build T from $Follow$ at optimal cost $O(2^m)$.

BuildT $(Follow, m)$
1. $T[0] \leftarrow 0^{m+1}$
2.   **For** $i \in 0 \ldots m$ **Do**
3.   **For** $j \in 0 \ldots 2^i - 1$ **Do** $T[2^i + j] \leftarrow Follow(i) \mid T[j]$
4.   **Return** T

Fig. 2. Construction of table T from Glushkov's variables. We use a numeric notation for the argument of T and use $Follow$ in bit mask form

3.4   A Search Algorithm

We present now the search algorithm based on the previous construction. Let us call $First$ and $Last$ the variables corresponding to the whole regular expression.

Our first step will be to set $Follow(0) = First$ for technical convenience. Second, we will add a self loop at state 0 which can be traversed by any $\sigma \in \Sigma$. This is because, for searching purposes, the NFA that *recognizes* a regular expression must be converted into one that *searches* the regular expression. This is achieved by appending Σ^* at its beginning, or which is the same, adding a self-loop as described. As, by Lemma 3, no arrow goes to state 0, it still holds that all the arrows leading to a state are labeled the same way (Lemma 1). Figure 3 shows the search algorithm.

Compared to Wu and Manber's algorithm [ ], ours has the advantage of needing $(m+1)(2^{m+1} + |\Sigma|)$ bits of space instead of their $m(2^{2m+1} + |\Sigma|)$ bits in the worst case (their best case is equal to our complexity). Just as they propose, we can split T horizontally to reduce space, so as to obtain $O(mn/\log s)$ time with $O(s)$ space. Compared to our previous algorithm [ ], the new one compares favorably against its $(m+1)2^{m+1}|\Sigma|$ bits of space. Therefore, our new algorithm should be always preferred over previous bit parallel algorithms.

With respect to a classical DFA implementation, its worst case is 2^{m+1} states, and it stores a table which for each state and each character stores the new state. This requires $(m+1)2^{m+1}|\Sigma|$ bits in the worst case. However, in the classical algorithm it is customary to build only the states that can actually be reached, which can be much less than all the 2^{m+1} possibilities.

We can do something similar, in the sense of filling only the reachable cells of T (yet, we cannot pack them consecutively as a classical DFA). Figure 4 shows the recursive construction of this table, which is invoked with $D = 0^m 1$, the initial state, and assumes that T is initialized with zeros and that B, $Follow$ and m are already computed.

```
Search(RE, T = t₁t₂ ... tₙ)
1.    Preprocessing
2.        (v_RE, m) ← Parse(RE) /* parse the regular expression */
3.        Glushkov_variables(v_RE,0) /* build the variables on the tree */
4.        Follow(0) ← 0ᵐ1 | First /* add initial self-loop */
5.        For σ ∈ Σ Do B[σ] ← B[σ] | 0ᵐ1
6.        T ← BuildT(Follow,m) /* build T table */
7.    Searching
8.        D ← 0ᵐ1 /* the initial state */
9.        For j ∈ 1 ... n Do
10.           If D & Last ≠ 0ᵐ⁺¹ Then report an occurrence ending at j − 1
11.           D ← T[D] & B[tⱼ] /* simulate transition */
```

Fig. 3. Glushkov-based bit-parallel search algorithm. We assume that **Parse** gives the syntax tree v_{RE} and the number of positions m in RE, and that **Glushkov_variables** builds B, $First$, $Last$, $Follow$ and $Empty$

```
BuildTrec (D)
1.    For i ∈ 0 ... m Do /* first build T[D] */
2.        If D & 0ᵐ⁻ⁱ10ⁱ ≠ 0ᵐ⁺¹ Then T[D] ← T[D] | Follow(i)
3.    For σ ∈ Σ Do
4.        If T[N & B[σ]] = 0ᵐ⁺¹ Then BuildTrec (N & B[σ])
```

Fig. 4. Recursive construction of table T. We fill only the reachable cells

4   Experimental Results

We compare in this section our approach against previous work. We use two different texts: an English one (writings of B. Franklin, filtered to lower-case) and a DNA sequence (h.influenzae). Both were replicated until obtaining 10 Mb.

A major problem when presenting experiments on regular expressions is that there is not a concept of "random" regular expression, so it is not possible to search, say, 1,000 random patterns. Lacking such good choice, we fixed a set of 10 patterns for English and 10 for DNA, which were selected to illustrate different interesting cases rather than more or less "probable" cases. Therefore, the goal is not to show what are the typical cases in practice but to show how the scheme behaves under different characteristics of the pattern.

The patterns are given in Table 1. We also show their number of letters, which is closely related to the size of the automata recognizing them. The period (.) in the patterns matches any character except the end of line (lines have approximately 70 characters). We also use $[c_1...c_k]$ (where c_i are characters) as a shorthand for $(c_1|...|c_k)$. Instead of a character c, a range $c^1\text{-}c^2$ can be specified to avoid enumerating all the letters between (and including) c^1 and c^2.

Table 1. The patterns used on DNA and English

No.	DNA Pattern	m	No.	English Pattern	m
1	AC((A|G)T)*A	6	1	benjamin|franklin	16
2	AGT(TGACAG)*A	10	2	benjamin|franklin|writing	23
3	(A(T|C)G)|((CG)*A)	7	3	[a-z][a-z0-9]*[a-z]	3
4	GTT|T|AG*	6	4	benj.*min	8
5	A(G|CT)*	4	5	[a-z][a-z][a-z][a-z][a-z]	5
6	((A|CG)*|(AC(T|G))*)AG	9	6	(benj.*min)|(fra.*lin)	15
7	AG(TC|G)*TA	7	7	ben(a|(j|a)*)min	9
8	[ACG][ACG][ACG][ACG][ACG][ACG]T	7	8	be.*ja.*in	8
9	TTTTTTTTTT[AG]	11	9	ben[jl]amin	8
10	AGT.*AGT	7	10	(be|fr)(nj|an)(am|kl)in	14

Our machine is a Sun UltraSparc-1 of 167 MHz, with 64 Mb of RAM, running Solaris 2.5.1. We measured CPU times in seconds, averaging 100 runs over the 10 Mb (the variance was very low).

We have tested the following forward scanning algorithms (the implementations are ours except otherwise stated). See the Introduction for detailed descriptions of previous work. **DFA** uses the classical deterministic automaton (we have not minimized the automaton). **Agrep** [,] uses bit parallelism on Thompson's construction (we forced it to use one table except for English pattern #2). **Ours (old)** is our previous forward algorithm of [ ] (it builds only the reachable states, just like **DFA**). **Ours (naive)** is our new algorithm building the whole table T with **BuildT**. **Ours (optim)** is our new algorithm where we build only the T mask for the reachable states, using **BuildTrec**. We left aside some algorithms which proved not competitive, at least for the sizes of the regular expressions we are considering: Thompson's [ ] and Myers' [ ]. This last one should be competitive for larger patterns.

The goal of showing two versions of our algorithm is as follows. Our normal algorithm builds the complete T_d table for all the 2^{m+1} possible combinations (reachable or not) of active and inactive states. It permits comparing directly against Agrep and to show that our technique is superior. Our optimized algorithm builds only the reachable states and it permits comparing against DFA (the classical algorithm) and our old algorithm. The disadvantage of our optimized algorithm is that it does not permit splitting the tables (neither does DFA), while our "naive" algorithm and that of Agrep do.

Table 2 shows the results on the different patterns, where we have separated preprocessing and search time. As it can be seen, our new algorithm (naive version) compares favorably in search time against Agrep, scanning (averaging over the 20 patterns) 16.0 Mb/sec versus about 13.2 Mb/sec of Agrep. It works quite well except on large patterns like the natural language pattern #2. Our optimized algorithm behaves well in those situations too, and compares favorably against the classical DFA algorithm and our old bit parallel algorithm, which scan the text at 14.4 and 14.6 Mb/sec, respectively. This means that our new algorithm is at least 10% faster than any alternative approach.

In all cases, searching larger expressions costs more, both in preprocessing and in search time because of locality of reference. Note that our optimized algorithm is sometimes worse than the naive one. This occurs when most states are reachable, in which case the naive algorithm fills all them without the overhead of the recursion. But this only happens when the preprocessing time is negligible.

Table 2. Search times in the form of $a + bn$, where a is the preprocessing time and b is the search time per megabyte, all in tenths of seconds

#	DFA	Agrep	Ours (old)	Ours (naive)	Ours (optim)
			DNA text		
1	$0.034 + 0.643n$	$0.104 + 0.756n$	$0.007 + 0.631n$	$0.009 + 0.584n$	$0.005 + 0.575n$
2	$0.006 + 0.624n$	$0.133 + 0.754n$	$0.011 + 0.630n$	$0.049 + 0.566n$	$0.000 + 0.576n$
3	$0.028 + 0.796n$	$0.095 + 0.758n$	$0.007 + 0.803n$	$0.007 + 0.759n$	$0.087 + 0.775n$
4	$0.025 + 0.883n$	$0.101 + 0.760n$	$0.008 + 0.865n$	$0.012 + 0.788n$	$0.029 + 0.807n$
5	$0.018 + 0.831n$	$0.089 + 0.757n$	$0.007 + 0.814n$	$0.005 + 0.755n$	$0.008 + 0.777n$
6	$0.007 + 0.658n$	$0.126 + 0.762n$	$0.008 + 0.652n$	$0.014 + 0.584n$	$0.004 + 0.592n$
7	$0.004 + 0.634n$	$0.104 + 0.750n$	$0.005 + 0.635n$	$0.015 + 0.571n$	$0.040 + 0.567n$
8	$0.004 + 0.646n$	$0.101 + 0.831n$	$0.012 + 0.638n$	$0.008 + 0.583n$	$0.071 + 0.582n$
9	$0.006 + 0.621n$	$0.096 + 0.694n$	$0.024 + 0.626n$	$0.005 + 0.568n$	$0.007 + 0.565n$
10	$0.038 + 0.639n$	$0.108 + 0.748n$	$0.018 + 0.645n$	$0.011 + 0.560n$	$0.036 + 0.562n$
			English text		
1	$0.010 + 0.633n$	$0.114 + 0.779n$	$0.006 + 0.633n$	$0.074 + 0.569n$	$0.009 + 0.563n$
2	$0.022 + 0.629n$	$0.112 + 1.583n$	$0.009 + 0.699n$	$20.61 + 0.575n$	$0.019 + 0.569n$
3	$0.003 + 0.932n$	$0.106 + 0.769n$	$0.050 + 0.897n$	$0.007 + 0.856n$	$0.022 + 0.898n$
4	$0.068 + 0.639n$	$0.100 + 0.755n$	$0.023 + 0.631n$	$0.008 + 0.567n$	$0.000 + 0.578n$
5	$0.009 + 0.879n$	$0.095 + 0.871n$	$0.063 + 0.727n$	$0.050 + 0.664n$	$0.000 + 0.684n$
6	$0.242 + 0.645n$	$1.494 + 0.775n$	$0.083 + 0.640n$	$0.043 + 0.569n$	$0.013 + 0.567n$
7	$0.007 + 0.631n$	$0.122 + 0.755n$	$0.006 + 0.625n$	$0.006 + 0.572n$	$0.001 + 0.578n$
8	$0.081 + 0.628n$	$0.103 + 0.755n$	$0.023 + 0.624n$	$0.048 + 0.562n$	$0.027 + 0.556n$
9	$0.000 + 0.627n$	$0.102 + 0.704n$	$0.008 + 0.626n$	$0.011 + 0.567n$	$0.002 + 0.565n$
10	$0.012 + 0.632n$	$0.774 + 0.789n$	$0.007 + 0.643n$	$0.018 + 0.567n$	$0.005 + 0.561n$

5   Conclusions

We have presented a new technique for compact DFA representation based on the properties of Glushkov's NFA construction, as opposed to the much better known Thompson's. As a result, we can represent the DFA using $(m + 1)(2^{m+1} + |\Sigma|)$ bits (where m is the number of normal characters in the pattern and Σ is the alphabet). This compares favorably against previous techniques which needed either $(m + 1)2^{m+1}|\Sigma|$ or $m(2^{2m+1} + |\Sigma|)$ bits.

The representation is quite practical. We are not only still able of searching in $O(n)$ time using the compact DFA, but thanks to more locality of reference we can search faster in practice than any previous approach, as we show experimentally.

Despite that we have presented a forward scan algorithm, our approach can be adapted to character skipping algorithms as well. For example, our algorithm presented in [ ] modified the automaton by reversing its arrows, making all the states initial and making the initial state final, so as to recognize reverse prefixes of the original language $L(RE)$. This algorithm is used to extend BNDM [ ] so as to obtain a fast character skipping algorithm for regular expression search.

Reversing the arrows means that the property that all arrows arriving to a state have the same label does not hold anymore (once we reverse the arrows, the result is not a Glushkov NFA). Rather, all the arrows *departing* from a state have now the same label. Once again, we can represent the DFA in a compact form by noting that $\delta(D, \sigma) = T\,[D\,\&B[\sigma]]$, where T is the deterministic *Follow* table of the reversed automaton and B is the character table of the original automaton. That is, we first keep the states of D that can originate arrows labeled by σ, and once they are obtained we find all the target states.

References

1. A. Aho, R. Sethi, and J. Ullman. *Compilers: Principles, Techniques and Tools*. Addison-Wesley, 1985. 1, 2
2. R. Baeza-Yates and G. Gonnet. A new approach to text searching. *CACM*, 35(10):74–82, October 1992. 2
3. G. Berry and R. Sethi. From regular expression to deterministic automata. *Theoretical Computer Science*, 48(1):117–126, 1986. 3, 4
4. A. Brüggemann-Klein. Regular expressions into finite automata. *Theoretical Computer Science*, 120(2):197–213, November 1993. 5
5. C.-H. Chang and R. Paige. From regular expression to DFA's using NFA's. In *Proceedings of the 3rd Annual Symposium on Combinatorial Pattern Matching*, LNCS v. 664, pages 90–110, 1992. 5
6. V.-M. Glushkov. The abstract theory of automata. *Russian Mathematical Surveys*, 16:1–53, 1961. 3, 4
7. E. Myers. A four-russian algorithm for regular expression pattern matching. *J. of the ACM*, 39(2):430–448, 1992. 2, 4, 6, 10
8. G. Navarro and M. Raffinot. Fast regular expression search. In *Proceedings of the 3rd Workshop on Algorithm Engineering*, LNCS v. 1668, pages 199–213, 1999. 3, 8, 10, 12
9. G. Navarro and M. Raffinot. Fast and flexible string matching by combining bit-parallelism and suffix automata. *ACM Journal of Experimental Algorithmics (JEA)*, 5(4), 2000. http://www.jea.acm.org/2000/NavarroString. 12
10. K. Thompson. Regular expression search algorithm. *CACM*, 11(6):419–422, 1968. 1, 2, 4, 10
11. B. Watson. *Taxonomies and Toolkits of Regular Language Algorithms*. Phd. dissertation, Eindhoven University of Technology, The Netherlands, 1995. 3
12. S. Wu and U. Manber. Agrep – a fast approximate pattern-matching tool. In *Proc. of USENIX Technical Conference*, pages 153–162, 1992. 2, 4, 10
13. S. Wu and U. Manber. Fast text searching allowing errors. *CACM*, 35(10):83–91, October 1992. 2, 4, 6, 8, 10

The Max-Shift Algorithm for Approximate String Matching

Costas S. Iliopoulos[1,2]*, Laurent Mouchard[3,4]**, and
Yoan J. Pinzon[1,2]***

[1] Dept. of Computer Science, King's College London
London WC2R 2LS, England
[2] School of Computing, Curtin University of Technology
GPO Box 1987 U, WA.
{csi,pinzon}@dcs.kcl.ac.uk
www.dcs.kcl.ac.uk/staff/csi, www.dcs.kcl.ac.uk/pg/pinzon
[3] ESA 6037: Dept. of Vegetal Physiology - ABISS, Université de Rouen
76821 Mont Saint Aignan Cedex, France
[4] School of Computing, Curtin University of Technology
GPO Box 1987 U, WA.
lm@dir.univ-rouen.fr
www.dir.univ-rouen.fr/~lm

Abstract. The approximate string matching problem is to find all locations which a pattern of length m matches a substring of a text of length n with at most k differences. The program *agrep* is a simple and practical bit-vector algorithm for this problem. In this paper we consider the following incremental version of the problem: given an appropriate encoding of a comparison between A and bB, can one compute the answer for A and B, and the answer for A and Bc with equal efficiency, where b and c are additional symbols? Here we present an elegant and very easy to implement bit-vector algorithm for answering these questions that requires only $O(n\lceil m/w \rceil)$ time, where n is the length of A, m is the length of B and w is the number of bits in a machine word. We also present an $O(nm\lceil h/w \rceil)$ algorithm for the *fixed-length approximate string matching problem*: given a text t, a pattern p and an integer h, compute the optimal alignment of all substrings of p of length h and a substring of t.

Keywords: String algorithms, approximate string matching, dynamic programming, edit-distance.

1 Introduction

The problem of finding substrings of a text similar to a given pattern has been intensively studied over the last twenty years and it is a central problem in a

---

* Partially supported by a Marie Curie fellowship and NATO, Wellcome and Royal Society grants.
** Partially supported by the C.N.R.S. Program "Génomes".
*** Partially supported by an ORS studentship and EPSRC Project GR/L92150.

G. Brodal et al. (Eds.): WAE 2001, LNCS 2141, pp. 13–25, 2001.
© Springer-Verlag Berlin Heidelberg 2001

wide range of applications: file comparison [ ], spelling correction [ ], information retrieval [ ], and searching for similarities among biosequences [, ,]. Given two strings A and B, we want to find an *alignment* between the two strings that exposes their similarity. An alignment is any pairing of symbols subject to the restriction that if lines were drawn between paired symbols as in the Figure 1 below, the lines would not cross. Scores are assigned to alignments according to the concept of similarity or difference required by the context of the application, and one seeks alignments of optimal score [ ].

One of the most common variants of the approximate string matching problem is that of finding substrings that match the pattern with at most k-differences. The first algorithm addressing exactly this problem is attributed to Sellers [ ]. This algorithm requires $O(nm)$ time, where n and m are the length of the text and of the query. Subsequently, this algorithm was refined to run in $O(kn)$ expected time [ ], and then to $O(kn)$ worst-case time, first with $O(n)$ space [ ], and later with $O(m^2)$ space [ ].

A new thread of practice-oriented results exploited the hardware word-level parallelism of bit-vector operations. In [ ], Baeza-Yates and Gonnet presented an $O(nm/w)$ algorithm for the exact matching case and an $O(nm \log k/w)$ algorithm for the k-mismatches problem, where w is the number of bits in a machine word. Wu and Manber [ ] showed an $O(nkm/w)$ algorithm for the k-differences problem. Furthermore, Wright [ ] presented an $O(n \log |\Sigma| m/w)$ bit-vector style algorithm, where $|\Sigma|$ is the size of alphabet for the pattern. Wu, Manber and Myers [ ] developed a particularly practical realization of the 4-Russians approach introduced by Masek and Paterson [ ]. Most recently, Baeza-Yates and Navarro [ ] have shown a $O(nkm/w)$ variation on the Wu/Manber algorithm, implying an $O(n)$ performance when $mk = O(w)$.

In this paper we consider the following incremental versions of the sequence comparison problem: given a solution for the comparison of A and B ($B = b\hat{B}$), can one incrementally compute a solution for A and $\hat{B}$? and for A and $\hat{B}c$? where b and c are additional symbols. By solution we mean some encoding of a relevant portion of the traditional dynamic programming matrix D computed by comparing A and B. D is an $(m + 1) \times (n + 1)$ matrix, where entry $D(i, j)$ is the best score for the problem of comparing $A[1..i]$ with $B[1..j]$. The data-dependencies in matrix D are such that it is easy to extend D to a matrix D' for A and Bc but it is quite difficult to extend D to a matrix D'' for A and $\hat{B}$. In essence, we are required to work against the "grain" of these data-dependencies. The further observation that matrices D and D' can differ on $O(nm)$ entries, suggests that the relationship between such adjacent problems is non-trivial. Here we present a bit-vector algorithm that answers the above queries in $O(n\lceil m/w \rceil)$ time. Landau, Myers and Schmidt [ ] demonstrated the power of efficient algorithms answering the above questions with a variety of applications to computational problems such as: "the longest common subsequence problem","the longest prefix approximate match problem", "approximate overlaps in the fragment assembly problem", "cyclic string comparison" and "text screen updating".

We also consider the *fixed-length approximate string matching problem*: given a text t, a pattern p and an integer h, compute the optimal alignment of all substrings of p of length h and a substring of t. This problem can be solved by computing $O(n)$ dynamic-programming matrices of size $O(nh)$ using any fast bit-vector algorithm for the approximate string matching such as Myers [ ] or Baeza-Yates and Navarro algorithm [,]. The total complexity of this approach will be $O(nm\lceil h/w \rceil)$ time and $O(nh)$ space. Here we present an algorithm with the same time/space complexity but using a new and simpler bit-wise technique which makes it much faster in practice. Our algorithm is independent of k and as such it can be used to compute blocks of dynamic programming matrix as the 4-Russians algorithm [ ]. Our algorithm is considerably simpler than the one presented by Landau, Myers and Schmidt [ ] and does not depend on the alphabet size, resulting in an improved performance.

The paper is organised as follows. In the next section we present some basic definitions. In Section 3 we present our main contribution (the MAX-SHIFT algorithm), and in Section 4 we describe an application and show how to use the MAX-SHIFT algorithm to solve it. Finally, in Section 5 we give our conclusions.

2   Basic Definitions

Consider the sequences $t_1 t_2 \dots t_r$ and $p_1 p_2 \dots p_r$ with $t_i, p_i \in \Sigma \cup \{\epsilon\}, i \in \{1..r\}$, where Σ is an *alphabet*, *i.e.* a set of symbols and ϵ is the empty string. If $t_i \neq p_i$, then we say that t_i *differs* from p_i. We distinguish among the following three types of differences:

1. [mismatch] A symbol of the first sequence corresponds to a different symbol of the second sequence, *i.e.* $t_i \neq p_i$.
2. [deletion] A symbol of the first sequence corresponds to "no symbol" of the second sequence, *i.e.* $t_i \neq \epsilon$ and $p_i = \epsilon$.
3. [insertion] A symbol of the second sequence corresponds to "no symbol" of the first sequence, *i.e.* $t_i = \epsilon$ and $p_i \neq \epsilon$.

Fig. 1. Types of differences: mismatch, deletion, insertion

As an example, Figure 1 shows a possible alignment between `"BADFECA"` and `"BCDEFCA"`. Positions 1, 3, 5, 7 and 8 are "matches". Positions 2, 4 and 6 are a "mismatch", a "deletion", and an "insertion", respectively. Without loss of generality, in the sequel we omit the empty string ϵ from the sequence of symbols in a string. Note that while for applications such as comparing protein sequences, the methods of scoring can involve arbitrary scores for symbol pairs and for gaps of unaligned symbols, in many other contexts simple unit cost schemes suffice.

Let $t = t_1 t_2 \dots t_n$ and $p = p_1 p_2 \dots p_m$ with $m \leq n$. We say that p occurs at position q of t with at most k differences (or equivalently, a *local alignment of* p *and* t *at position* q *with at most* k *differences*), if $t_q \dots t_r$, for some $r > q$, can be transformed into p by performing at most k operations (insertions, deletions, substitutions). Furthermore, we will use function $\delta(p, t)$ to denote the minimum number of operations required to transform p into t.

```
           1  2  3  4  5  6  7  8  9  10 11 12 13
String t:  A  B  C  B  B  A  D  F  E  ε  F  E  A
                    |        |     |                 Alignment #1
String p:           B  C  ε  D  ε  E  F  A  F

           1  2  3  4  5  6  7  8  9  10 11 12 13
String t:  A  B  C  B  B  A  D  F  E  ε  F  E  A
              |  |           |     |                 Alignment #2
String p:     B  C  ε  ε  ε  D  ε  E  F  A  F

           1  2  3  4  5  6  7  8  9  10 11 12 13
String t:  A  B  C  B  B  A  D  F  E  ε  F  E  A
                       |     |     |                 Alignment #3
String p:              B  C  D  ε  E  F  A  F

           1  2  3  4  5  6  7  8  9  10 11 12 13
String t:  A  B  C  B  B  A  D  ε  F  E  F  E  A
                       |     |     |     |           Alignment #4
String p:              B  C  D  E  F  A  F
```

Fig. 2. String searching with k-differences

For $t =$ "ABCBBADFEFEA" and $p =$ "BCDEFAF", Figure 2 shows four different alignments of p into t occurring at positions 4,2,5 and 5 with 6, 7, 5 and 3 differences, respectively. The alignment (or alignments) with the minimum number of differences is called *optimal alignment*.

We define D' as the *incremental* matrix containing the minimum number of differences (best scores) of the alignments of all substrings of p of length h and any contiguous substring of t. Table 1 shows the matrix D' for $t = p =$ "GGGTCTA" and h=3. For instance, $D'(5, 6)$ is the score for the best alignment between "GTC" ($p3 \dots p5$) and any substring of t ending at position 6.

3 The Max-Shift Algorithm

One can obtain a straightforward $O(nmh)$ algorithm for computing matrix D' by constructing matrices $D^{(s)}[1..h, 1..n]$, $1 \leq s \leq m - h + 1$, where $D^{(s)}(i, j)$ is the minimum number of differences between the prefix of length i of the pattern $p_s \dots p_{s+h-1}$ and any contiguous substring of the text ending at t_j; its computation can be based on the Dynamic-Programming procedure presented in [8]. We can obtain D' by collating $D^{(1)}$ and the last row of the $D^{(s)}$, $2 \leq s \leq m - h + 1$. (see Fig. 3).

Table 1. Matrix D' for $t = p =$ "GGGTCTA" and $h = 3$

		t							
		0	1	2	3	4	5	6	7
		ϵ	G	G	G	T	C	T	A
0	ϵ	0	0	0	0	0	0	0	0
1	G	1	0	0	0	1	1	1	1
2	G	2	1	0	0	1	2	2	2
3	G	3	2	1	0	1	2	3	3
4	T	3	2	1	1	0	1	2	3
5	C	3	2	2	2	1	0	1	2
6	T	3	3	3	3	2	1	0	1
7	A	3	3	3	3	2	2	1	0

(left label: p)

Here we will make use of word-level parallelism in order to compute matrix D' more efficiently. The algorithm is based on the $O(1)$-time computation of each $D'(i, j)$ by using bit-vector operations under the assumption that $h \leq w$, where w is the number of bits in a machine word or $O(\lceil h/w \rceil)$ time for the general case; thus on a "64-bit computer word" machine one can obtain a speed-up factor of 64.

We define the bit-vector $B(i, j) = b_\ell...b_1$, where $b_r = 1$, $r \in \{1...\ell\}$, $\ell < 2h$, if and only if there is an alignment of a contiguous substring of the text $t_q...t_j$ (for some $1 \leq q < j$) and $p_{i-h+1}...p_i$ with $D'(i, j)$ differences such that

- the leftmost $r - 1$ pairs of the alignment have $\Sigma_\ell^{\ell-r+2} b_j$ differences in total.
- the r-th pair of the alignment (from left to right) is a difference: a deletion in the pattern, an insertion in the text or a replacement.

Otherwise we set $b_r = 0$. In other words, $B(i, j)$ holds the binary encoding of the path in D' necessary to obtain an optimal alignment at (i, j).

Table 2 shows an example of the bit-vector matrix computation for $t = p =$ "GGGTCTA" and $h=3$. For instance, $B(5, 7)$ represents the binary encoding of the alignment between "GTC" ($p_3...p_5$) and "CTA" ($t_5...t_7$). $B(5, 7) = 101$ because we need to replace "G" with "C" (first 1 in 101), match "T" with T (middle 0 in 101) and replace "C" with "A" (last 1 in 101).

Given the constraint $h \leq w$, each entry of the matrix B can be computed in constant time using "bit-vector" operations (e.g. "shift"-type operations). The maintenance of the bit-vector is done via the following operations:

- SHL shifts bits one position to the left. i.e. $\text{SHL}(b_\ell...b_1) = b_\ell...b_1 0$
- SHLC same as SHL but truncates the leftmost bit. i.e. $\text{SHLC}(b_\ell ... b_1) = b_{\ell-1} ... b_1 0$
- MINMAX(x, y, z) returns r one of the integers $\{x, y, z\}$ with the least number of 1's (bits set on). If there is more than one candidate then returns the maximum (when they are viewed as decimal integers). Notice that the maximum is also the number with 1's occurring as leftmost as possible.

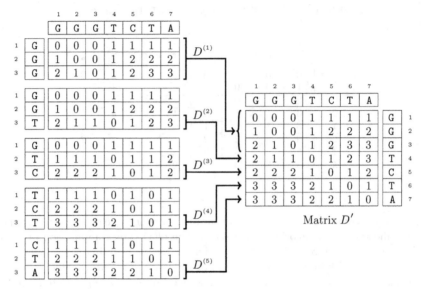

Fig. 3. Naive computation of matrix D' for $t = p =$ "GGGTCTA" and $h = 3$

- LAST returns the leftmost bit. *i.e.* LAST($b_\ell...b_1$) = b_ℓ
- OR corresponds to the bitwise-or operation.

SHL, SHLC, LAST and OR operations can be implemented in $O(1)$ time. MINMAX also can be implemented in $O(1)$ time if we keeping a counter of the number of bits set on along with the bit-vector.

The algorithm in Fig. 4 implements the above operations and ideas to compute matrix $B[0..n, 0..n]$.

Fig. 5 illustrates the computation of $B(3, 2)$ for our previous example, using the MAX-SHIFT algorithm (Fig. 4). Since $i \leq h$ ($i = 3, h = 3$) in line 6, we use the straightforward dynamic programing formula in line 8. To compute $B(3, 2)$ we need the bit-vectors $B(3, 1)$, $B(2, 1)$ and $B(2, 2)$. These values are shown in the previous stage. To get to stage a we simply shift (to the left) by one bit all the bit-vectors ($B(3, 1)$, $B(2, 1)$ and $B(2, 2)$). To pass from stage a to stage b, we need to OR $B(3, 1)$ and $B(2, 2)$ with 1, and $B(2, 1)$ with $\delta(p_3, t_2) = \delta("G", "G") = 0$. To get to the final stage we need to decide between $B(2, 1)$ and $B(2, 2)$ (both with one bit set on) and we select $B(2, 1)$ because $B(2, 1) = 4 > B(2, 2) = 1$ (when viewed as decimal integers).

Fig. 6 shows the computation of $B(7, 5)$. In this case, $i > h$ ($i = 7, h = 3$) and therefore $B(7, 5)$ will be computed using the formula in line 11. Stages a and stage b are as before but using SHLC instead of SHL for $B(6, 4)$ and $B(6, 5)$. That is why the last bit for $B(6, 4)$ and $B(6, 5)$ has been dropped before applying the SHL function. Now, from stage b we find the bit-vector with less number of ones set on (it can be $B(6, 4)$ or $B(6, 5)$ both with two 1's set on)

Table 2. Bit-vector matrix B for $t = p =$ "GGGTCTA" and $h = 3$

			0	1	2	3	4	5	6	7
			ϵ	G	G	G	T	C	T	A
	0	ϵ								
	1	G	1	0	0	0	1	1	1	1
	2	G	11	10	00	00	01	11	11	11
	3	G	111	110	100	000	001	011	111	111
p	4	T	111	101	001	001	000	0001	110	111
	5	C	111	011	011	011	001	000	0001	101
	6	T	111	111	111	111	**110**	**001**	000	0001
	7	A	111	111	111	111	**101**	**101**	001	000

MAX-SHIFT(t, p, n, m, h)

```
 1  begin
 2      ▷ Initialization
 3      B[0..m, 0] ← max(i, h) 1's;  B[0, 0..n] ← 0
 4      for i ← 1 until m do
 5        for j ← 1 until n do
 6          if i ≤ h then
 7            ▷ Straightforward-DP
 8            B(i, j) ← MINMAX{SHL(B(i − 1, j)) OR 1, SHL(B(i, j − 1)) OR 1,
                                SHL(B(i − 1, j − 1)) OR δ(t_i, t_j)}
 9          else
10            ▷ Max-Shift
11            B(i, j) ← MINMAX{SHLC(B(i − 1, j)) OR 1, SHL(B(i, j − 1)) OR 1,
                                SHLC(B(i − 1, j − 1)) OR δ(t_i, t_j)}
12  end
```

Fig. 4. MAX-SHIFT algorithm

---

and maximum decimal value (5 for $B(6, 4)$ and 3 for $B(6, 5)$). So the "winner" is $B(6, 4)$ (101) and we assign it to $B(7, 5)$.

Assume that the bit-vector $B[0..n, 0..n]$ is given. We can use B as an input for the INCREMENTAL-DP algorithm (see Fig. 7) to compute matrix D', however, matrix D' can be obtained concurrently with the computation of the matrix B.

The function ONES(v) returns the number of 1's (bits set on) in bit-vector v.

Theorem 1. *Given the text t, the pattern p and an integer h, the MAX-SHIFT algorithm correctly computes the matrix B in $O(nm\lceil h/w \rceil)$ units of time.*

Proof. The computation of $B(i, j)$ for all $1 \leq i, j \leq m$, in line 8, is done using straightforward dynamic programing (see [ ]) and thus their values are correct.

Let us consider the computation of $B(i, j)$ for some $i > h$. The value of $D'(i, j)$ (number of 1's in $B(i, j)$) denotes the minimum number of differences in an optimal alignment of a contiguous substring of the text $t_q \ldots t_j$ (for some

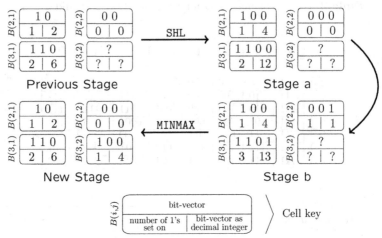

Fig. 5. Illustration of $B(3,2)$ computation for $t = p =$ "GGGTCTA" and $h = 3$

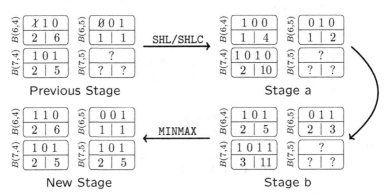

Fig. 6. Illustration of $B(7,5)$ computation for $t = p =$ "GGGTCTA" and $h = 3$

$1 \le q < j$) and $p_{i-h+1} \ldots p_i$; in that alignment p_i can be either to the right of t_j or to the left of t_j or aligned with t_j. It is clear that for $q < j$:

$$D'(i, j) = \delta(p_{i-h+1} \ldots p_i, t_q \ldots t_j) \tag{1}$$

Now, we will consider all three cases. In the case that the symbol p_i is aligned to the right of t_j, for some $q' < j$, we have

$$\delta(p_{i-h+1} \ldots p_i, t_q \ldots t_j) = \delta(p_{i-h+1} \ldots p_{i-1}, t_{q'} \ldots t_j) + 1 \tag{2}$$

Let us consider, for some $\hat{q} < j$

$$D'(i-1, j) = \delta(p_{i-h} p_{i-h+1} \ldots p_{i-1}, t_{\hat{q}} \ldots t_j) \tag{3}$$

INCREMENTAL-DP(t, p, n, m, h, B)
1 **begin**
2   **for** $i \leftarrow 1$ **until** m **do**
3   **for** $j \leftarrow 1$ **until** n **do**
4       $D'(i, j) \leftarrow$ ONES($B(i, j)$)
5 **end**

Fig. 7. INCREMENTAL-DP algorithm

$$\delta(p_{i-h}p_{i-h+1} \ldots p_{i-1}, t_{\hat{q}} \ldots t_j) = \delta(p_{i-h}, t_{\hat{q}}) + \delta(p_{i-h+1} \ldots p_{i-1}, t_{q'} \ldots t_j) \quad (4)$$

Note that

$$\delta(p_{i-h}, t_{\hat{q}}) = \text{LAST}(B(i, j-1)) \quad (5)$$

From equations 1-5 follows that

$$D'(i, j) = D'(i-1, j) + 1 - \text{LAST}(B(i, j-1)) \quad (6)$$

Now we consider the subcase $\hat{q} \le q'$. In this case p_{i-h} is either paired with $t_{\hat{q}}$ or with ϵ in an optimal alignment with score $\delta(p_{i-h}p_{i-h+1} \ldots p_{i-1}, t_{\hat{q}} \ldots t_j)$. Thus we have either

$$\delta(p_{i-h}p_{i-h+1} \ldots p_{i-1}, t_{\hat{q}} \ldots t_j) = \delta(p_{i-h}, t_{\hat{q}}) + \delta(p_{i-h+1} \ldots p_{i-1}, t_{\hat{q}-1} \ldots t_j) \quad (7)$$

or

$$\delta(p_{i-h}p_{i-h+1} \ldots p_{i-1}, t_{\hat{q}} \ldots t_j) = \delta(p_{i-h}, \epsilon) + \delta(p_{i-h+1} \ldots p_{i-1}, t_{\hat{q}} \ldots t_j) \quad (8)$$

It is not difficult to see that

$$\begin{aligned}
\delta(p_{i-h+1} \ldots p_{i-1}, t_{q'} \ldots t_j) &= \delta(p_{i-h+1} \ldots p_{i-1}, t_{\hat{q}-1} \ldots t_j) \\
&= \delta(p_{i-h+1} \ldots p_{i-1}, t_{\hat{q}} \ldots t_j)
\end{aligned} \quad (9)$$

From 1-3, 5, 7 or 8 and 9, we also derive 6 in this subcase.

In the case that the symbol p_i is aligned to the left of t_j (as above), we have

$$\delta(p_{i-h+1} \ldots p_i, t_q \ldots t_j) = \delta(p_{i-h+1} \ldots p_i, t_{q'} \ldots t_{j-1}) + 1 = D'(i, j-1) + 1$$

which implies that

$$D'(i, j) = D'(i, j-1) + 1 \quad (10)$$

In the case that the symbol p_i is aligned with t_j (as above), we have

$$\delta(p_{i-h+1} \ldots p_i, t_q \ldots t_j) = \delta(p_{i-h+1} \ldots p_{i-1}, t_{q'} \ldots t_{j-1}) + \delta(p_i, t_j) \quad (11)$$

In a similar manner as in 2-5 we can show that

$$\delta(p_{i-h+1} \ldots p_{i-1}, t_{q'} \ldots t_{j-1}) = D'(i-1, j-1) - \text{LAST}(B(i-1, j-1)) \quad (12)$$

and from 11-12 follows that

$$D'(i,j) = D(i-1, j-1) + \delta(p_i, t_j) - \texttt{LAST}(B(i-1, j-1)) \qquad (13)$$

Equations 6, 10 and 13 are equivalent to line 11 of the MAX-SHIFT algorithm and thus the algorithm's correctness follows.

The worst-case running time of the MAX-SHIFT algorithm can easily be shown to be $O(nm\lceil h/w \rceil)$. □

Theorem 2. *The matrix D' can be computed in $O(nm\lceil h/w \rceil)$ units of time.*

Proof. The computation of matrix D' can be done concurrently with the computation of the matrix B. □

Hence, this algorithm runs in $O(nm)$ under the assumption that $h \leq w$ and its space complexity can be reduced to $O(n)$ by nothing that each row of B depends only on its immediately preceding row.

3.1   Experimental Results

We implemented the MAX-SHIFT algorithm using g++ 2.81 and compared its performance against Myers algorithm [ ] and Baeza-Yates and Navarro algorithm [,] ("BYN"). Of course, we can not use these algorithms directly because they solve a different problem, namely, the string pattern matching with k differences problem. Nevertheless, they can be adapted very easily using the naive way presented in Fig. 3. These adapted algorithms compute a row of the D'-matrix in $O(n\lceil h/w \rceil)$ and the whole matrix in $O(nm\lceil h/w \rceil)$. Hence, the complexity of these algorithm are identical to that of the MAX-SHIFT. However, although they all use bit-vector operations, the bit-wise techniques used to develop them are complete different. For instance, the MAX-SHIFT algorithm is the first algorithm that uses a bit-wise technique based on the encoding of the alignments. Another important feature in favor of the MAX-SHIFT algorithm is that it does not use any kind of preprocessed information. All these algorithms were implemented so that they can cope with the general case when $h > w$ (known as the *unbounded* model or the *unrestricted* model). The implementation of the unbounded model for the MAX-SHIFT was simpler because we do not need to deal with bit carries and there are not special cases to be consider. The results are shown in Fig. 8. All trials were run on a SUN Ultra Enterprise 300MHz running Solaris Unix with a 32-bits word size. We used random text drawn from a uniformly distributed alphabets of size 26 and assumed $t = p$, $h = 10$ and $k = 3$. The MAX-SHIFT was always faster for the values used during this experiment.

4   Some Application

In this section we show how to apply the MAX-SHIFT algorithm to solve the *cyclic string comparison problem*. Consider the text $t = t_1 t_2 ... t_n$ and the pattern

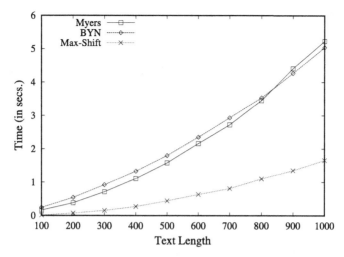

Fig. 8. Timing curves for the Myers algorithm ("Myers"), Baeza-Yates and Navarro algorithm ("BYN") and the Max-Shift algorithm ("Max-Shift") for $t = p$, $h = 10$ and $k = 3$

$p = p_1 p_2 ... p_m$. Let $cycle(t_1 t_2 ... t_n) = t_2 ... t_n t_1$, and let $cycle^r(t)$ be the result of applying $cycle$ exactly r times. *i.e.* $cycle^r(t) = \hat{t}$ where $\hat{t}_j = t \bmod (j+r-1,n)+1$ $\forall j \in \{1..n\}$. The *cyclic string comparison problem* is to determine the integers r and s such that $d = \delta(cycle^r(t), cycle^s(p))$ is minimal. It is quite easy to see that if the minimum is obtained for $cycle^r(t)$ and $cycle^s(p)$ then by simply cyclically shifting an alignment achieving this minimum, one obtains an equally good alignment between t and $cycle^s(p)$ for some s. Hence, the problem reduces to find s such that $\delta(t, cycle^s(p))$ is minimal. This problem was first introduce by Mathias Maes [ ] and he gave an $O(mn \log m)$ time complexity algorithm. Here, we present a simple and practical $O(mn)$ (for $m < w$) algorithm based on the Max-Shift algorithm presented in the previous section (see also [ ]).

Consider comparing t and $\overline{p} = p \cdot p$ (p concatenated with itself). First of all, we use Max-Shift$(t, \overline{p}, m)$ to compute the matrix of bit-vector $B[1..2m, 1..n]$. Let k be the row of the cell with less number of 1's among $B[m..2m-1, n]$. We say that $s = k - m$ gives the best alignment between t and $cycle^s(p)$.

For example, for $t=$ "DCABCD" and $p=$ "CDBA", table 3 shows the bit-vector matrix B for t and $\overline{p}=$ "CDBACDBA". We are then looking for the row with minimum score (lest number of 1's) among $B[4..7, 6]$. We can see that row 6 has score 1 and is the minimum. Hence, $s=6-4=2$ and we conclude that the best alignment between t and $cycle^s(p)$ is for $s = 2$ ($\delta($"DCABCD", "BACD"$) = 1$) and it is the minimum possible.

Table 3. The bit-vector matrix B for $t =$ "DCABCD", $p =$ "CDBACDB" and $h = 4$

		0	1	2	3	4	5	6	
		ϵ	B	C	A	B	C	D	
0	ϵ								
1	C	1	1	0	1	1	0	1	
2	D	11	10	01	01	11	01	00	
3	B	111	101	101	011	010	011	001	
4	A	1111	1011	1011	1010	0101	0101	0011	$s = 0$
5	C	1111	0111	0110	0101	0101	1010	1011	$s = 1$
6	D	1111	1110	1101	1101	1011	0101	**0100**	$s = 2$
7	B	1111	1101	1101	1011	1010	1011	1001	$s = 3$

5   Conclusion

Here we presented a new algorithm for some variations of approximate string pattern matching problem. The main advantages of the new algorithm over algorithms such as those in [3] and [11] are: it is faster in practice, simpler, easy to implement, does not require preprocessing (*i.e.* does not use/store look up tables) and does not depend on k (the number of differences).

References

1. R. A. Baeza-Yates and G. H. Gonnet, A new approach to text searching, *CACM*, Vol 35, (1992), pp. 74–82. 14
2. R. A. Baeza-Yates and G. Navarro, A faster algorithm for approximate string matching, in *Proceedings of the 7th Symposium on Combinatorial Pattern Matching*, LNCS, Vol. 1075, Springer-Verlag, New York, (1996), pp. 1-23. 14, 15, 22
3. R. A. Baeza-Yates and G. Navarro, Analysis for algorithm engineering: Improving an algorithm for approximate pattern matching. Unpublished manuscript. 15, 22, 24
4. Z. Galil and K. Park, An improved algorithm for approximate string matching, *SIAM Journal on Computing*, **19** (1990), pp. 989–999. 14
5. P. A. Hall and G. R. Dowling, Approximate string matching, *Computing Surveys*, Vol 12, (1980), pp. 381–402. 14
6. J. W. Hunt and T. G. Szymanski, An algorithm for differential file comparison, *Comm. of the ACM*, Vol 20, (1977), pp. 350–353. 14
7. G. M. Landau, E. Myers and J. P. Schmidt, Incremental string comparison, *SIAM Journal on Computing* 27, 2 (1998), 557-582. 14, 15, 23
8. G. M. Landau and U. Vishkin, Fast string matching with k differences, *Journal of Computer and Systems Sciences*, **37** (1988), pp. 63–78. 14, 16, 19
9. M. Maes, On a cyclic string-to-string correction problem, in *Info. Proc. Lett.*, Vol. 35, (1990), pp. 73–78. 23
10. W. J. Masek and M. S. Paterson, A Fast algorithm for computing string edit distances. in *J. Comput. Sy. Sci.*, Vol. 20, (1980), pp. 18–31. 14

11. E. W. Myers , A Fast Bit-Vector Algorithm for Approximate String Matching Based on Dynamic Progamming, in *Journal of the ACM* 46,3 (1999) pp. 395–415. 15, 22, 24

12. S. B. Needleman and C. D. Wunsch, A general method applicable to the search for similarities in the amino acid sequence of the two proteins, in *J. of Mol. Bio.*, Vol 48 (1970), pp. 443–453. 14

13. P. H. Seller, The theory and computation of evolutionary distances: Pattern recognition, in *Journal of Algorithms*, Vol 1, (1980), pp. 359–373. 14

14. T. F. Smith and M. S. Waterman, Identification of common molecular subsequences, in *Journal of Molecular Biology*, Vol 147, No. 2 (1981), pp. 195-197. 14

15. E. Ukkonen, Finding approximate patterns in strings, in *J. of Algorithms*, Vol 6, (1985), pp. 132-137. 14

16. R. A. Wangner and M. J. Fischer, The string-to-string correction problem, in *J. of the ACM*, Vol 21, No. 1 (1974), pp. 168-173. 14

17. A. H. Wright, Approximate string matching using within-word parallelism. in *Soft. Pract. Exper.*, Vol 24, (1994), pp. 337-362. 14

18. S. Wu and U. Manber, Fast text searching allowing errors, *CACM*, Vol 35, (1992), pp. 83–91. 14

19. S. Wu, U. Manber and G. Myers, A subquadratic algorithm for approximate limited expression matching, in *Algorithmica*, Vol. 15, (1996), pp. 50–67. 14, 15

Fractal Matrix Multiplication: A Case Study on Portability of Cache Performance

Gianfranco Bilardi[1], Paolo D'Alberto[2], and Alex Nicolau[2]

[1] Dipartimento di Elettronica e Informatica, Università di Padova, Italy
bilardi@dei.unipd.it[* * *]
[2] Information and Computer Science, University of California at Irvine
{paolo,nicolau}@ics.uci.edu[†]

Abstract. The practical portability of a simple version of matrix multiplication is demonstrated. The multiplication algorithm is designed to exploit maximal and predictable locality at all levels of the memory hierarchy, with no *a priori* knowledge of the specific memory system organization for any particular machine. By both simulations and execution on a number of platforms, we show that memory hierarchies portability does not sacrifice floating point performance; indeed, it is always a significant fraction of peak and, at least on one machine, is higher than the tuned routines by both ATLAS and vendor. The results are obtained by careful algorithm engineering, which combines a number of known as well as novel implementation ideas. This effort can be viewed as an experimental case study, complementary to the theoretical investigations on portability of cache performance begun by Bilardi and Peserico.

1 Introduction

The ratio between main memory access time and processor clock cycle has been continuously increasing, up to values of a few hundreds nowadays. The increase in Instruction Level Parallelism (ILP) has been a significant feature: current CPUs can issue four/six instructions per cycle and the cost of a memory access is an increasingly high toll on overall performance of super-scalar/VLIW processors. The architectural response has been an increase in the size and number of caches, with a second level being available on most machines, and a third level becoming now popular. The memory hierarchy helps performance only to the extent to which the computation exhibits data and code locality. The necessary amount of locality becomes greater with steeper hierarchies, an issue that algorithm design and compiler optimization increasingly need to take into account. A number of studies have begun to explore these issues. An early paper by Aggarwal, Alpern, Chandra, and Snir [ ] introduced the Hierarchical Memory Model (HMM) of computation, as a basis to design and evaluate memory efficient algorithms (extended in [,]). In this model, the time to access a location x is a function

---

* * * This work was supported, in part, by CNR and MURST of Italy
† Supported by AMRM DABT63-98-C-0045

G. Brodal et al. (Eds.): WAE 2001, LNCS 2141, pp. 26–38, 2001.
© Springer-Verlag Berlin Heidelberg 2001

$f(x)$; the authors observe that optimal algorithms are achieved for a wide family of functions f. More recently, similar results have been obtained for a different model, with automatically managed caches [ ]. The optimality is established by deriving a lower bound to the access complexity $Q(S)$, i.e., to the number of accesses that necessarily miss any given set of S memory locations. Lower bounds techniques were pioneered in [ ] and recently extended in [,]; these techniques are crucial to establish the existence of portable implementations for some algorithms, such as matrix multiplication. The question whether arbitrary computations admit optimally portable implementations has been investigated in [,]. Even though the answer is generally negative, the computations that admit portable implementations do include relevant classes such as linear algebra kernels ([,   ]).

This work focuses on matrix multiplications algorithms with complexity $O(n^3)$ (rather than $O(n^{\log_2 7})$ [ ] or $O(n^{2.376})$ [ ]) investigating the impact on performance of data layout, latency hiding, register allocation, instruction scheduling, instruction parallelism, (e.g., [,   ,   ,   ,   ])[1] and their interdependences. The interdependence between tiling and sizes of caches is probably the most investigated [,   ,   ,   ,   ,   ,   ]. For example, vendor libraries (such as BLAS from SGI and SUN) exploit their knowledge of the destination platform and determine very efficient routines, but non optimally portable across different platforms. Automatically tuned packages (see [,   ] matrix multiply and [ ] FFT) measure *machine parameters* by interactive tests and then produce machine tuned code. This approach achieves optimal performance and *portability* at the level of package, rather than the actual application code. Another approach, called *auto-blocking*, has the potential to yield portable performance for the individual code. Informally, one can think of a tile whose size is not determined by any *a priori* information but arises automatically from a recursive decomposition of the problem. This approach has been advocated in [ ], with applications to LAPACK, and its asymptotic optimality is discussed in [ ]. Our fractal algorithms belong to this framework. Recursion-based algorithms often exploit various features of non-standard layouts, *recursive layouts* ([,   ,   ,   ,   ,   ,   ]). Conversion from and to standard (i.e., row-major and column-major) layouts introduces $O(n^2)$ overheads[2]. Recursive algorithms are often based on power of two matrixes (with padding, overlapping, or peeling) because of closure properties of the decomposition and a simple index computation. In this paper, we use a non-padded layout for arbitrary square matrices, thus saving space and maintaining the conceptual simplicity of the algorithm, while developing an approach to burst the recursion and save index computations. Register allocation and instruction scheduling are still bottlenecks ([,]); for recursive algorithms the problem is worse because no compiler is capable of unfolding the calls in or-

---

[1] See [, , , , ,] for more general locality approaches suitable at compile time and used for linear algebra kernels.

[2] The overheads are negligible, except for matrices small enough for the n^2/n^3 ratio to be insignificant, or large enough to require disk access.

der to expose larger sets of operations to aggressive optimizations. We propose a pruning of the recursion tree to circumvent this problem.

Our approach, hereafter *fractal* approach, combines a number of known ideas and techniques as well as some novel ones to achieve the following results.

1) There exists a matrix multiplication implementation for modern ILP machines achieving excellent, portable cache performance, and we show it through simulations of 7 different machines. 2) The overall performance (FLOPS) is very good in practice, and we show it by comparison with the upper bound implied by peak and performance of the best known code (Automatically Tuned Linear Algebra Software, ATLAS, [39]). 3) While the main motivation to develop the fractal approach was provided by the goal of portability, at least on some machines such as the R5000 IP32, the fractal approach yields the fastest known algorithms. Among the techniques we have developed, those in Sections 2.2 and 2.3 lead to efficient implementations of recursive procedures. They are especially worth mentioning because they are likely to be applicable to many other hierarchy-oriented codes. In fact, it can be argued with some generality that recursive code is naturally more conducive to express temporal locality than code written in the form of nested loops. Numerical stability is not considered in this paper (Lemma 2.4.1 resp. 3.4 in [24] resp. [27]).

2   Fractal Algorithms for Matrix Multiplication

We use the following recursive layout of an $m \times n$ matrix A into a one-dimensional array $\mathbf{a}$ of size mn. If $m = 1$, then $\mathbf{a}[h] = a_{0h}$, for $h = 0, 1, \ldots, n-1$. If $n = 1$, then $\mathbf{a}[h] = a_{h0}$, for $h = 0, 1, \ldots, m-1$. Otherwise, $\mathbf{a}$ is the concatenation of the layouts of the blocks A_0, A_1, A_2, and A_3 of the following *balanced* decomposition. $A_0 = \{a_{ij} : 0 \leq i < \lceil m/2 \rceil, 0 \leq j < \lceil n/2 \rceil\}$, $A_1 = \{a_{ij} : 0 \leq i < \lceil m/2 \rceil, \lceil n/2 \rceil \leq j < n\}$, $A_2 = \{a_{ij} : \lceil m/2 \rceil \leq I < m, 0 \leq j < \lceil n/2 \rceil\}$ and $A_3 = \{a_{ij} : \lceil m/2 \rceil \leq i < m, \lceil n/2 \rceil \leq j < n\}$. A $m \times n$ matrix is said *near square* when $|n - m| \leq 1$. If A is a near-square matrix, so are the blocks A_0, A_1, A_2, and A_3 of its balanced decomposition. Indeed, a straightforward case analysis ($m = n - 1, n, n + 1$ and m even or odd) shows that, if $|n - m| \leq 1$ and $S = \{\lfloor m/2 \rfloor, \lceil m/2 \rceil, \lfloor n/2 \rfloor, \lceil n/2 \rceil\}$, then $\max(S) - \min(S) \leq 1$. The fractal layout just defined can be viewed as a generalization of the Z-Morton layout for square matrixes [12], [20] or as a special case of the Quad-Tree [19] layout.

We introduce now the fractal algorithms, a class of procedures all variants of a common scheme, for the operation of matrix multiply-and-add (MADD) $C = C + AB$, also denoted $C+ = AB$. For near square matrices, the *fractal scheme* to perform $C+ = AB$ is recursively defined as follows, with reference to the above balanced decomposition.

fractal(A, B, C)

- If $|A| = |B| = 1$, then $C = C + A * B$ (all matrices being scalar).
- Else, execute - in any serial order - the calls **fractal**(A', B', C') for

$$(A', B', C') \in \{(A_0, B_0, C_0), (A_1, B_2, C_0), (A_0, B_1, C_1), (A_1, B_3, C_1),$$
$$(A_2, B_0, C_2), (A_3, B_2, C_2), (A_2, B_1, C_3), (A_3, B_3, C_3)\}$$

Of particular interest, from the perspective of temporal locality, are those orderings where there is always a sub-matrix in common between consecutive calls, which increases data reuse. The problem of finding such orderings can be formulated by defining an undirected graph. The vertices correspond to the 8 recursive calls in the fractal scheme. The edges join calls that share exactly one sub-matrix (observe that no two calls share more than one sub-matrix). This graph is easily recognized to be a 3D binary cube. An ordering that maximizes data reuse corresponds to an Hamiltonian path in this cube (See Fig. 1).

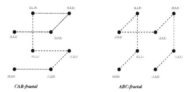

Fig. 1. The cube of calls of the fractal scheme: the Hamiltonian path defining CAB-fractal and ABC-fractal

Even when restricting our attention to Hamiltonian orderings, there are many possibilities. The exact performance of each of them depends on the specific structure and policy of the machine cache(s) in a way too complex to evaluate analytically and too time consuming to evaluate experimentally. In this paper, we shall focus on two orderings: one reducing write misses and one reducing read misses. We call **CAB-fractal** the algorithm obtained from the fractal scheme when the recursive calls are executed in the following order: (A_0, B_0, C_0), (A_1, B_2, C_0), (A_1, B_3, C_1), (A_0, B_1, C_1), (A_2, B_1, C_3), (A_3, B_3, C_3), (A_3, B_2, C_2), (A_2, B_0, C_2). The label "CAB" underlines the fact that sub-matrix sharing between consecutive calls is maximum for C (4 cases), medium for A (2 cases), and minimum for B (1 case). It is reasonable to expect that CAB-fractal will tend to better reduce write misses, since C is the matrix being written. In a similar vein, but with a stress on reducing read misses, we consider the algorithm **ABC-fractal** obtained from the fractal scheme when the recursive calls are executed in the following order: (A_0, B_0, C_0), (A_0, B_1, C_1), (A_2, B_1, C_3), (A_2, B_0, C_2), (A_3, B_2, C_2), (A_3, B_3, C_3), (A_1, B_3, C_1), (A_1, B_2, C_0).

2.1   Cache Performance

Fractal multiplication algorithms can be implemented with respect to any memory layout of the matrices. For an ideal fully associative cache with least recently used replacement policy (LRU) and with cache lines holding exactly one matrix entry, the layout is immaterial to performance. The fractal approach exploits temporal locality for any cache independently of its size s (in matrix entries).

Indeed, consider the case when at the highest level of recursion all calls use matrix blocks that fit in cache simultaneously. Approximately, the matrix blocks are of size $s/3$. Each call load will cause about s misses. Each call computes up to $(\sqrt{s/3})^3 = s\sqrt{s}/3\sqrt{3}$ scalar MADDs. The ratio misses per FLOP is estimated as $\mu = (3\sqrt{3}(/(2\sqrt{s}) \approx 2.6/\sqrt{s}$. (This is within a constant factor of optimal, Corollary 6.2 [ ].)

For a real machine, the above analysis needs to be refined, keeping into account the effects of cache-line length ℓ (in matrix entries) and a low degree of associativity. Here, the fractal layout, which stores relevant matrix blocks in contiguous memory locations, takes full advantage of cache-line effects and has no self interference for blocks that fit in cache. The misses per flop is estimated as $\mu = 2.6\gamma/\ell\sqrt{s}$, where γ accounts for cross interference between different matrices and other fine effects not captured by our analysis. In general, for a given fractal algorithm, γ will depend on matrix size (n), relative fractal arrays positions in memory, cache associativity and, sometimes, register allocation. When interference is negligible, we can expect $\gamma \approx 1$.

2.2   The Structure of the Call Tree

Pursuing efficient implementations for the fractal algorithms we face the usual performance drawbacks of recursion: overheads and poor register utilization (due to lack of code exposure to the compiler). To circumvent such drawbacks, we carefully study the structure of the call tree.

Definition 1. *Given a fractal algorithm $\mathcal{A}$, its call tree $T = (V, E)$ w.r.t. input (A, B, C) is an ordered, rooted tree defined as follows. V contains one node for each call. The root of T corresponds to the main call* **fractal**(A,B,C). *The ordered children $v_1, v_2, \ldots, v_8$ of an internal node v correspond to the calls made by v in order of execution.*

If A is $m \times n$ and B is $n \times p$, we shall say that the input is of *type* $< m, n, p >$. If one among m, n, and p is zero, then we shall say that the type is *empty* and use also the notation $< \emptyset >$. The structure of T is uniquely determined by type of the root. We focus on square matrices, i.e. type $< n, n, n >$ for which the tree has depth $\lceil \log n \rceil + 1$ and it has $8^{\lceil \log n \rceil}$ leaves. n^3 leaves have type $< 1, 1, 1 >$ and correspond (from left to right) to the n^3 MADDs of the algorithm. The remaining leaves have empty type. Internal nodes are essentially responsible for performing the problem decomposition; their specific computation depends on the way matrices are represented. An internal node has typically eight nonempty children, except when its type has at least one components equal to 1, e.g., $< 2, 1, 1 >$ or $< 2, 2, 1 >$, in which the non empty children are 2 and 4, respectively. While the call tree has about n^3 nodes, most of them have the same type. To deal with this issue systematically, we introduce the concept of *type DAG*. Given a fractal algorithm $\mathcal{A}$, an input type $< m, n, p >$, and the corresponding call tree $T = (V, E)$, the *call type DAG* $D = (U, F)$ is a DAG, where the arcs with the same source are ordered, such that: 1) U contains exactly one node for each type occurring in T, the node corresponding to $< m, n, p >$ is called the *root* of D; 2) F contains, for each $u \in U$, the ordered set of arcs $(u, w_1), \ldots, (u, w_8)$, where $w_1, \ldots, w_8$ are the types of the (ordered) children of any node in T with type u. See Figure 2 for an example. Next, we study the size of the call-type DAG D for the case of square matrix multiplication. We begin by showing that there are at most 8 types of input for the calls of a given level of recursion.

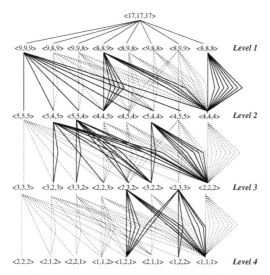

Fig. 2. Example of call-type DAG for Matrix Multiplication $< 17, 17, 17 >$

Proposition 1. *For any integers $n \geq 1$ and $d \geq 0$, let n_d be defined inductively as $n_0 = n$ and $n_{d+1} = \lceil n_d/2 \rceil$. Also, for any integer $q \geq 1$, define the set of types $Y(q) = \{< r, s, t >: r, s, t \in \{q, q - 1\} \}$. Then, in the call tree corresponding to a type $< n, n, n >$, the type of each call-tree node at distance d from the root belongs to the set $Y(n_d)$, for $d = 0, 1, \ldots, \lceil \log n \rceil$.*

Proof. The statement trivially holds for $d = 0$ (the root), since $< n, n, n > \in Y(n) = Y(n_0)$. Assume now inductively that the statement holds for a given level d. From the closure property of the balance decomposition and the recursive decomposition of the algorithm, it follows that all matrix blocks at level $d+1$ have dimensions between $\lfloor (n_d - 1)/2 \rfloor$ and $\lceil n_d/2 \rceil$. From the identity $\lfloor (n_d - 1)/2 \rfloor = \lceil n_d/2 \rceil - 1$, we have that all types at level $d + 1$ belong to $Y(\lceil n_d/2 \rceil) = Y(n_{d+1})$.

Now, we can give an accurate size estimate of the call-type DAG.

Proposition 2. *Let n be of the form $n = 2^k s$, with s odd. Let $D = (U, F)$ be the call-type DAG corresponding to input type $< n, n, n >$. Then, $|U| \leq k + 1 + 8(\lceil \log n \rceil - k)$.*

Proof. It is easy to see that, at level $d = 0, 1, \ldots, k$ of call tree nodes have type $< n_d, n_d, n_d >$, with $n_d = n/2^d$. For each of the remaining ($\lceil \log n \rceil - k$) levels, there are at most 8 types per level, according to Proposition 1.

Thus, we always have $|U| = O(\log n)$, with $|U| = \log n + 1$ when n is a power of two, with $|U| \approx 8\lceil \log n \rceil$ when n is odd, and with $|U|$ somewhere in between for general n.

2.3   Bursting the Recursion

If v is an internal node of the call tree, the corresponding call receives as input a triple of blocks of A, B, and C, and produces as output the input for each child call. When matrices A, B, and C are *fractally* represented by the corresponding one-dimensional arrays a, b, and c, the input triple is uniquely determined by the type $< r, s, t >$ and by the initial positions i, j, and k of the blocks in their respective arrays. Specifically, the block of A is stored in $a[i, \ldots, i + rs - 1]$, the block of B is stored in $b[j, \ldots, j + st - 1]$, and the block of C is stored in $c[k, \ldots, k + rt - 1]$. The call at v is then responsible for the computation of the type and initial position of the sub-blocks processed by the children. For example, for the A-block $r \times s$ starting at i, the four sub-blocks have respective dimensions $\lceil r/2 \rceil \times \lceil s/2 \rceil$, $\lceil r/2 \rceil \times \lfloor s/2 \rfloor$, $\lfloor r/2 \rfloor \times \lceil s/2 \rceil$, and $\lfloor r/2 \rfloor \times \lfloor s/2 \rfloor$. They also have respective starting points i_0, i_1, i_2, and i_3, of the form $i_h = i + \Delta i_h$, where: $\Delta i_0 = 0$, $\Delta i_1 = \lceil r/2 \rceil \lceil s/2 \rceil$, $\Delta i_2 = \lceil r/2 \rceil s$, $\Delta i_3 = \Delta i_2 + \lfloor r/2 \rfloor \lceil s/2 \rceil$. In a similar way, one can define the analogous quantities $j_h = j + \Delta j_h$ for the sub-blocks of B and $k_h = k + \Delta k_h$ for the sub-blocks of C, for $h = 0, 1, 2, 3$. During the recursion and in any node of the call tree, every Δ value is computed twice. Hence, a straightforward implementation of the fractal algorithm is bound to be rather inefficient. Two avenues can be followed, separately or in combination. First, rather than executing the full call tree down to the n^3 leaves of type $< 1, 1, 1 >$, one can execute a pruned version of the tree. This approach reduces the recursion overheads and the straight-line coded leaves are amenable to aggressive register allocation, a subject of the next section. Second, the integer operations are mostly the same for all calls. Hence, these operations can be performed in a preprocessing phase, storing the results in an auxiliary data structure built around the call-type DAG D, to be accessed during the actual processing of the matrices. Counting the number of instructions per node, we can see a reduction of 30%.

3   Register Issues

The impact of register management on overall performance is captured by the number ρ of memory (load or store) operations per floating point operation, required by a given assembly code. In a single-pipeline machine with at most one FP or memory operation per cycle, $1/(1 + \rho)$ is an upper limit to the achievable fraction of FP peak performance. The fraction lowers to $1/(1 + 2\rho)$ for machines where MADD is available as a single-cycle instruction. For machines with parallel pipes, say 1 load/store pipe every f FP pipes, an upper limit to the achievable fraction of FP peak performance becomes $\max(1, f\rho)$, so that memory instructions are not a bottleneck as long as $\rho \leq 1/f$. In this section, we explore two techniques which, for the typical number of registers of current RISC processors, lead to values of ρ approximately in the range $1/4$ to $1/2$. The general approach consists in stopping the recursion at some point and formulating the corresponding leaf computation as a straight-line code. All matrix entries are copied into a set of scalar variables, whose number R is chosen so that any reasonable compiler will permanently keep these variables in registers (*scalarization*). For a given R, the goal is then to choose where to stop the recursion and how to sequence the operations so as to minimize ρ, i.e., to minimize the number of assignments to and from scalar variables.

We investigated and implemented two different scalar replacements: *Fractal Sequence* "inspired" by [ ] and *C-tiling sequence* inspired by [ ] (see [,] for a full description).

4 Experimental Results

We have studied experimentally both the cache behavior of fractal algorithms, in terms of misses, and the overall performance, in terms of running time.

4.1 Cache Misses

The results of this section are based on simulations performed (on an SPARC Ultra 5) using the *Shade* software package for Solaris, of Sun Microsystems. Codes are compiled for the SPARC Ultra2 processor architecture (V8+, no MADD operation available) and then simulated for various cache configurations, chosen to correspond to those of a number of commercial machines. Thus when we refer, say, to the R5000 IP32, we are really simulating a ultra2 CPU with the memory hierarchy of the R5000 IP32.

In fractal codes, (i) the recursion is stopped when the size of the leaves is strictly smaller than problem $< 32, 32, 32 >$; (ii) the recursive layout is stopped when a sub-matrix is strictly smaller than 32×32; (iii) the leaves are implemented with C-tiling register assignment using $R = 24$ variables for scalarization (this leaves the compiler 8 of the 32 registers to buffer multiplication outputs before they are accumulated into C-entries). The leaves are compiled with cc WorkShop 4.2 and linked statically (as suggested in [ ]). The recursive algorithms, i.e. ABC-Fractal and CAB-Fractal, are compiled with gcc 2.95.1.

We have also simulated the code for ATLAS DGEMM obtained by installation of the package on the Ultra 5 architecture. This is used as another term of reference, and generally fractal has fewer misses. However, it would be unfair to regard this as a competitive comparison with ATLAS, which is meant to be efficient by adapting to the varying cache configuration. We have simulated 7 different cache configurations (Table 1); we use the notation: I= Instruction

Table 1. Summary of simulated configurations

Simulated	Conf.	Size (Bytes/s)	Line (Bytes,ℓ)	Assoc./ WritePol.	$\mu(1000)/$ $\gamma(1000)$
SPARC 1	U1	64KB / 8K	16B / 2	1 / through	2.65e-2 / 1.84
SPARC 5	I1	16KB	16B	1 /	
	D1	8KB / 1K	16B / 2	1 / through	5.96e-2 / 1.47
Ultra 5	I1	16KB	32B	2 /	
	D1	16KB / 2K	32B / 4	1 / through	2.51e-2 / 1.75
	U2	2MB / 256K	64B / 8	1 / back	1.05e-3 / 1.66
R5000 IP32	I1	32KB	32B	2 / back	
	D1	32KB / 4K	32B / 4	2 / back	1.06e-2 / 1.04
	U2	512KB / 64K	32B / 4	1 / back	3.61e-3 / 1.42
Pentium II	I1	16KB	32B	1 /	
	D1	16KB / 2K	32B / 4	1 / through	2.50e-2 / 1.74
	U2	512KB / 64K	32B / 4	1 / back	3.98e-3 / 1.57
HAL Station	I1	128KB	128B	4 / back	
	D1	128KB / 16K	128B / 16	4 / back	2.65e-3 / 2.09
ALPHA 21164	I1	8KB	32B	1 /	
	D1	8KB / 1K	32B / 4	1 / through	3.75e-2 / 1.85
	U2	96KB / 12K	32B / 4	3 / back	5.81e-3 / 0.99

cache, D=Data cache, and U=Unified cache. We have measured the number $\mu(n)$ of misses per flop and compared it against the value of the estimator (Section 2.1) $\mu(n) = 2.6\gamma(n)/(\ell\sqrt{s})$, where s and ℓ are the number of (64 bit) words in the cache and in one line, respectively, and where we expect values of $\gamma(n)$ not much greater than one. In Table 1, we have reported the value of $\mu(1000)$ measured for CAB-fractal and the corresponding value of $\gamma(1000)$ (last column). More detailed simulation results are given in [ ]. We can see that γ is generally between 1 and 2; thus, our estimator gives a reasonably accurate prediction of cache performance. This performance is consistently good on the various configurations, indicating efficient portability. For completeness, in [ ], we have also reported simulation results for code misses: although these misses do increase due to the comparatively large size of the leaf procedures, they remain negligible with respect to data misses.

4.2   MFLOPS

While portability of cache performance is desirable, it is important to explore the extent to which it can be combined with optimizations of CPU performance. We have tested the fractal approach on four different processors listed in Table 2. We always use the same code for the recursive decomposition (which is essentially responsible for cache behavior). We vary the code for the leaves, to adapt the number of scalar variables R to the processor: $R = 24$ for Ultra 5, $R = 8$ for Pentium II, and $R = 32$ for SGI R5K IP32 and HAL Station. We compare the MFLOPS of fractal algorithms in double precision with peak performance and with the performance of ATALS-DGEMM, if available. Fractal achieves performances comparable to those of ATLAS, being at most 2 times slower on PentiumII (which is not a RISC) and a little faster on SGI R5K. Since no special adaptation to the processor has been performed on the fractal codes, except for the number of scalar variables, we conclude that the portability of cache performance can be combined with overall performance. More detailed running time results are reported in [ ]

Table 2. Processor Configurations

Processor	Ultra 2i (Ultra 5)	PentiumII	R5000 (IP32)	HAL Station
Registers Structure	32 64-bit register file	8 80-bit stack file	32 64-bit register file	32 64-bit register file
Multiplier Adder	distinct	distinct	single FU	single FU
FP Lat.(Cycles)	3	8	2	4
Peak (MFLOPS)	666	400	360	200
Peak of CAB-Fr. / matrix size	$425 / 444 \times 444$	$187 / 400 \times 400$	$133 / 504 \times 504$	$168 / 512 \times 512$
Peak of ATLAS / matrix size	$455 / 220 \times 220$	$318 / 848 \times 848$	$113 /$ unknown	not available

5 Conclusions

In this paper, we have developed a careful study of matrix multiplication implementations, showing that suitable algorithms can efficiently exploit the cache hierarchy without taking cache parameters into account, thus ensuring portability of cache performance. Clearly, performance itself does depend on cache parameters and we have provided a reasonable estimator for it. We have also experimentally shown that, with a careful implementation of recursion, high performance is achievable. We hope the present study will motivate extension in various directions, both in terms of results and in terms of techniques. In [ ], we have already used the fractal multiplication codes and recursive code optimizations of this paper to obtain implementation of other linear algebra algorithms, such as those for LU decomposition of [ ], with overall performance higher than other multiplication-based algorithms.

References

1. A. Aggarwal, B. Alpern, A. K. Chandra and M. Snir: A model for hierarchical memory. Proc. of 19th Annual ACM Symposium on the Theory of Computing, New York, 1987,305-314. 26
2. A. Aggarwal, A. K. Chandra and M. Snir: Hierarchical memory with block transfer. 1987 IEEE.

3. B. Alpern, L. Carter, E. Feig and T. Selker: The uniform memory hierarchy model of computation. In *Algorithmica*, vol. 12, (1994), 72-129. 26

4. U. Banerjee, R. Eigenmann, A. Nicolau and D. Padua: Automatic program parallelization. Proceedings of the IEEE vol 81, n.2 Feb. 1993. 27

5. G. Bilardi, P. D'Alberto, and A. Nicolau: Fractal Matrix Multiplication: a Case Study on Portability of Cache Performance, *University of California at Irvine*, ICS TR#00-21, 2000. 33, 34, 35

6. G. Bilardi and F. P. Preparata: Processor-time tradeoffs under bounded-speed message propagation. Part II: lower bounds. Theory of Computing Systems, Vol. 32, 531-559, 1999. 27

7. G. Bilardi, E. Peserico: An Approach toward an Analytical Characterization of Locality and its Portability. *IWIA 2000, International Workshop on Innovative Architectures*, Maui, Hawai, January 2001. 27

8. G. Bilardi, E. Peserico: A Characterization of Temporal Locality and its Portability Across Memory Hierarchies. *ICALP 2001, International Colloquium on Automata, Languages, and Programming*, Crete, July 2001. 27

9. G. Bilardi, A. Pietracaprina, and P. D'Alberto: On the space and access complexity of computation DAGs. 26th Workshop on Graph-Theoretic Concepts in Computer Science, Konstanz, Germany, June 2000. 27

10. J. Bilmes, Krste Asanovic, C. Chin and J. Demmel: Optimizing matrix multiply using PHiPAC: a portable, high-performance, Ansi C coding methodology. International Conference on Supercomputing, July 1997. 27

11. S. Carr and K. Kennedy: Compiler blockability of numerical algorithms. Proceedings of Supercomputing Nov 1992, pg.114-124. 27

12. S. Chatterjee, V. V. Jain, A. R. Lebeck and S. Mundhra: Nonlinear array layouts for hierarchical memory systems. Proc. of ACM international Conference on Supercomputing, Rhodes,Greece, June 1999. 27, 28

13. S. Chatterjee, A. R. Lebeck, P. K. Patnala and M. Thottethodi: Recursive array layout and fast parallel matrix multiplication. Proc. 11-th ACM SIGPLAN, June 1999. 27

14. D. Coppersmith and S. Winograd: Matrix multiplication via arithmetic progression. In Poceedings of 9th annual ACM Symposium on Theory of Computing pag.1-6, 1987. 27

15. P. D'Alberto, G. Bilardi and A. Nicolau: Fractal LU-decomposition with partial pivoting. Manuscript. 35

16. M. J. Dayde and I. S. Duff: A blocked implementation of level 3 BLAS for RISC processors. TR_PA_96_06, available on line http://www.cerfacs.fr/algor reports/TR_PA_96_06.ps.gz Apr. 6 1996 27

17. N. Eiron, M. Rodeh and I. Steinwarts: Matrix multiplication: a case study of algorithm engineering. Proceedings WAE'98, Saarbrücken, Germany, Aug.20-22, 1998 27

18. Engineering and Scientific Subroutine Library. http://www.rs6000.ibm.com/ resource/aix_resource/sp_books/essl/ 27

19. P. Flajolet, G. Gonnet, C. Puech and J. M. Robson: The analysis of multidimentional searching in Quad-Tree. Proceeding of the second Annual ACM-SIAM symposium on Discrete Algorithms, San Francisco, 1991, pag.100-109. 27, 28

20. J. D. Frens and D. S. Wise: Auto-blocking matrix-multiplication or tracking BLAS3 performance from source code. Proc. 1997 ACM Symp. on Principles and Practice of Parallel Programming, SIGPLAN Not. 32, 7 (July 1997), 206–216. 27, 28, 33

21. M. Frigo and S. G. Johnson: The fastest Fourier transform in the west. MIT-LCS-TR-728 Massachusetts Institute of technology, Sep. 11 1997. 27

22. M. Frigo, C. E. Leiserson, H. Prokop and S. Ramachandran: Cache-oblivious algorithms. Proc. 40th Annual Symposium on Foundations of Computer Science, (1999). 27

23. E. D. Granston, W. Jalby and O. Teman: To copy or not to copy: a compile-time technique for assessing when data copying should be used to eliminate cache conflicts. Proceedings of Supercomputing Nov 1993, pg.410-419. 27

24. G. H. Golub and C. F. van Loan: Matrix computations. Johns Hopkins editor 3-rd edition. 28

25. F. G. Gustavson: Recursion leads to automatic variable blocking for dense linear algebra algorithms. Journal of Research and Development Volume 41, Number 6, November 1997. 27

26. F. Gustavson, A. Henriksson, I. Jonsson, P. Ling, and B. Kagstrom: Recursive blocked data formats and BLAS's for dense linear algebra algorithms. In B. Kagstrom et al (eds), Applied Parallel Computing. Large Scale Scientific and Industrial Problems, PARA'98 Proceedings. Lecture Notes in Computing Science, No. 1541, p. 195-206, Springer Verlag, 1998. 27

27. N. J. Higham: Accuracy and stability of numerical algorithms ed. SIAM 1996 28

28. Hong Jia-Wei and T. H. Kung: I/O complexity :The Red-Blue pebble game. Proc.of the 13th Ann. ACM Symposium on Theory of Computing Oct.1981,326-333. 27, 30

29. B. Kågström, P. Ling and C. Van Loan: Algorithm 784: GEMM-based level 3 BLAS: portability and optimization issues. ACM transactions on Mathematical Software, Vol24, No.3, Sept.1998, pages 303-316 27

30. B. Kågström, P. Ling and C. Van Loan: GEMM-based level 3 BLAS: high-performance model implementations and performance evaluation benchmark. ACM transactions on Mathematical Software, Vol24, No.3, Sept.1998, pages 268-302. 27

31. M. Lam, E. Rothberg and M. Wolfe: The cache performance and optimizations of blocked algorithms. Proceedings of the fourth international conference on architectural support for programming languages and operating system, Apr.1991,pg. 63-74. 27

32. S. S. Muchnick: Advanced compiler design implementation. Morgan Kaufman 27

33. P. D'Alberto: Performance Evaluation of Data Locality Exploitation. Techincal Report UBLCS-2000-9. Department of Computer Science, University of Bologna. 33

34. P. R. Panda, H. Nakamura, N. D. Dutt and A. Nicolau: Improving cache performance through tiling and data alignment. Solving Irregularly Structured Problems in Parallel Lecture Notes in Computer Science, Springer-Verlag 1997. 27

35. John E. Savage: Space-Time tradeoff in memory hierarchies. Technical report Oct 19, 1993. 26

36. V. Strassen: Gaussian elimination is not optimal. Numerische Mathematik 14(3):354-356, 1969. 27

37. S. Toledo: Locality of reference in LU decomposition with partial pivoting. SIAM J.Matrix Anal. Appl. Vol.18, No. 4, pp.1065-1081, Oct.1997 35

38. M. Thottethodi, S. Chatterjee and A. R. Lebeck: Tuning Strassen's matrix multiplication for memory efficiency. Proc. SC98, Orlando,FL, nov.1998 (http://www.supercomp.org/sc98). 27

39. R. C. Whaley and J. J. Dongarra: Automatically Tuned Linear Algebra Software. http://www.netlib.org/atlas/index.html 27, 28, 33

40. D. S. Wise: Undulant-block elimination and integer-preserving matrix inversion. Technical Report 418 Computer Science Department Indiana University August 1995 27
41. M. Wolfe: More iteration space tiling. Proceedings of Supercomputing, Nov.1989, pg. 655-665. 27
42. M. Wolfe and M. Lam: A Data locality optimizing algorithm. Proceedings of the ACM SIGPLAN'91 conference on programming Language Design and Implementation, Toronto, Ontario,Canada,June 26-28, 1991. 27
43. M. Wolfe: High performance compilers for parallel computing. Addison-Wesley Pub.Co.1995 27

Experiences with the Design and Implementation of Space-Efficient Deques

Jyrki Katajainen and Bjarke Buur Mortensen

Department of Computing, University of Copenhagen
Universitetsparken 1, DK-2100 Copenhagen East, Denmark
{jyrki, rodaz}@diku.dk
http://www.diku.dk/research-groups/performance-engineering/

Abstract. A new realization of a space-efficient deque is presented. The data structure is constructed from three singly resizable arrays, each of which is a blockwise-allocated pile (a heap without the order property). The data structure is easily explainable provided that one knows the classical heap concept. All core deque operations are performed in $O(1)$ worst-case time. Also, general modifying operations are provided which run in $O(\sqrt{n})$ time if the structure contains n elements. Experiences with an implementation of the data structure show that, compared to an existing library implementation, the constants for some of the operations are unfavourably high, whereas others show improved running times.

1 Introduction

A *deque* (*double-ended queue*) is a data type that represents a sequence which can grow and shrink at both ends efficiently. In addition, a deque supports random access to any element given its *index*. Insertion and erasure of elements in the middle of the sequence are also possible, but these should not be expected to perform as efficiently as the other operations. A deque is one of the most important components of the C++ standard library; sometimes it is even recommended to be used as a replacement for an array or a vector (see, e.g., [ ]).

Let X be a deque, n an index, p a valid iterator, q a valid dereferenceable iterator, and r a reference to an element. Of all the deque operations four are fundamental:

operation	effect
X.begin()	returns a random access iterator referring to the first element of X
X.end()	returns a random access iterator referring to the one-past-the-end element of X
X.insert(p, r)	inserts a copy of element referred to by r into X just before p
X.erase(q)	erases the element referred to by q from X

We call the insert and erase operations collectively the *modifying operations*. The semantics of the *sequence operations*, as they are called in the C++ standard, can be defined as follows:

G. Brodal et al. (Eds.): WAE 2001, LNCS 2141, pp. 39–50, 2001.
© Springer-Verlag Berlin Heidelberg 2001

operation	operational semantics
X[n]	*(X.begin() + n) (no bounds checking)
X.at(n)	*(X.begin() + n) (bounds-checked access)
X.front()	*(X.begin())
X.back()	*(--(X.end()))
X.push_front(r)	X.insert(X.begin(), r)
X.pop_front()	X.erase(X.begin())
X.push_back(r)	X.insert(X.end(), r)
X.pop_back()	X.erase(--(X.end()))

For a more complete description of all deque operations, we refer to the C++ standard [, Clause 23], to a textbook on C++, e.g., [ ], or to a textbook on the Standard Template Library (STL), e.g., [ ].

In this paper we report our experiences with the design and implementation of a deque which is space-efficient, supports fast sequence operations, and has relatively fast modifying operations. Our implementation is part of the Copenhagen STL which is an open-source library under development at the University of Copenhagen. The purpose of the Copenhagen STL project is to design alternative/enhanced versions of individual STL components using standard performance-engineering techniques. For further details, we refer to the Copenhagen STL website [ ].

The C++ standard states several requirements for the complexity of the operations, exception safety, and iterator validity. Here we focus on the time- and space-efficiency of the operations. According to the C++ standard all sequence operations should take $O(1)$ time in the worst case. By *time* we mean the sum of operations made on the elements manipulated, on iterators, and on any objects of the built-in types. Insertion of a single element into a deque is allowed to take time linear in the minimum of the number of elements between the beginning of the deque and the insertion point and the number of elements between the insertion point and the end of the deque. Similarly, erasure of a single element is allowed to take time linear in the minimum of the number of elements before the erased element and the number of elements after the erased element.

In the Silicon Graphics Inc. SGI) implementation of the STL [ ], a deque is realized using a number of data blocks of fixed size and an index block storing pointers to the beginning of the data blocks. Only the first and the last data block can be non-full, whereas all the other data blocks are full. Adding a new element at either end is done by inserting it into the first/last data block. If the relevant block is full, a new data block is allocated, the given element is put there, and a pointer to the new block is stored in the index block. If the index block is full, another larger index block is allocated and the pointers to the data blocks are moved there. Since the size of the index block is increased by a constant factor, the cost of the index block copying can be amortized over the push operations. Hence, the push operations are supported in $O(1)$ amortized time and all the other sequence operations in $O(1)$ worst-case time. Thus this realization is not fully compliant with the C++ standard. Also, the space allocated for the index block is never freed so the amount of extra space used is not necessarily proportional to the number of elements stored.

Recently, Brodnik et al. citeBCDMS99a announced the existence of a deque which performs the sequence operations in $O(1)$ worst-case time and which requires never more than $O(\sqrt{n})$ extra space (measured in elements and in objects of the built-in types) if the deque stores n elements. After reading their conference paper, we decided to include their deque realization in the Copenhagen STL. For the implementation details, they referred to their technical report [ ]. After reading the report, we realized that some implementation details were missing; we could fill in the missing details, but the implementation got quite complicated. The results of this first study are reported in [ ]. The main motivation of this first study was to understand the time/space tradeoff better in this context. Since the results were a bit unsatisfactory, we decided to design the new space-efficient data structure from scratch and test its competitiveness with SGI's deque.

The new design is described in Sections 2–5. For the sequence operations, our data structure gives the same time and space guarantees as the proposal of Brodnik et al. citeBCDMS99b. In addition, using the ideas of Goodrich and Kloss II [ ] we can provide modifying operations that run in $O(\sqrt{n})$ time. Our solution is based on an efficient implementation of a resizable array, i.e., a structure supporting efficient inserts and erases only at one end, which is similar to that presented by Brodnik et al. [,]. However, after observing that "deques cannot be efficiently implemented in the worst case with two stacks, unlike the case of queues", they use another paradigm for realizing a deque. While their observation is correct, we show that a deque can be realized quite easily using three resizable arrays. One can see our solution as a slight modification of the standard "two stacks" technique relying on global rebuilding [ ]. All in all, our data structure is easily explainable which was one of the design criteria of Brodnik et al.

The experimental results are reported in Section 6. Compared to SGI's deque, for our implementation shrink operations at the ends may be considerably slower, grow operations at the ends are only a bit slower, access operations are a bit faster, and modifying operations are an order of magnitude faster.

2   Levelwise-Allocated Piles

A *heap*, as defined by Williams [ ], is a data structure with the following four properties:

Shape property: It is a left-complete binary tree, i.e., a tree which is obtained from a complete binary tree by removing some of its rightmost leaves.

Capacity property: Each node of the tree stores one element of a given type.

Representation property: The tree is represented in an array $\mathbf{a}[0 \mathbin{.\!.} n)$ by storing the element at the root of the tree at entry 0, the elements at its children at entries 1 and 2, and so on.

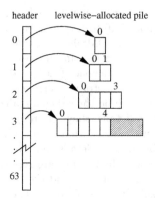

Fig. 1. A levelwise-allocated pile storing 12 elements

Order property: Assuming that we are given an ordering on the set of elements, for each branch node the element stored there is no smaller than the element stored at any children at that node.

Our data structure is based on a heap but for us the order property is irrelevant. For the sake of clarity, we call the data structure having only the shape, capacity, and representation properties a *static pile*.

The main drawback of a static pile is that its size n must be known beforehand. To allow the structure to grow and shrink at the back end, we allocate space for it levelwise and store only those levels that are not empty. We call this stem of the data structure a *levelwise-allocated pile*. We also need a separate array, called here the *header*, for storing the pointers to the beginning of each level of the pile. Theoretically, the size of this header is $\lceil \log_2(n+1) \rceil$, but a fixed array of size, say 64, will be sufficient for all practical purposes. The data structure is illustrated in Figure 1. Observe that element $\mathbf{a}[k]$, $k \in \{0, 1, \ldots, n-1\}$, has index $k - 2^{\lfloor \log_2(k+1) \rfloor} + 1$ at level $\lfloor \log_2(k+1) \rfloor$.

The origin of the levelwise-allocated pile is unclear. The first author of this paper gave the implementation of a levelwise-allocated heap as a programming exercise for his students in May 1998, but the idea is apparently older. Bojesen [ ] used the idea in the implementation of dynamic heaps in his heaplab. According to his experiments the practical performance of a levelwise-allocated heap is almost the same as that of the static heap when used in Williams' heapsort [ ].

If many consecutive grow and shrink operations are performed at a level boundary, it might happen that the memory for a level is repeatedly allocated and deallocated. We can assume that both of these memory-allocation operations require constant time, but in practice the constant is high (see [, Appendix 3]). To amortize the memory-allocation costs, we do not free the space reserved by the highest level h until all the elements from level $h-1$ have been erased. Also, it is appropriate to allocate the space for the first few levels (8 in our actual implementation) statically so that the extra costs caused by memory allocation can be avoided altogether for small piles.

For a data structure storing n elements, the space allocated for elements is never larger than $4n+O(1)$. Additionally, the extra space for $O(\log_2 n)$ pointers is needed by the header. If we ignore the costs caused by the dynamization of the header — as pointed out in practice there are no costs — a levelwise-allocated pile provides the same operations equally efficiently as a static pile. In addition, the grow and shrink operations at the back end are possible in $O(1)$ worst-case time. For instance, to locate an element only a few arithmetic operations are needed for determining its level and its position at that level; thereafter only two memory accesses are needed. To determine the level, at which element $\mathbf{a}[k]$ lies, we have to compute $\lfloor \log_2(k+1) \rfloor$. Since the computation of the whole-number logarithm of a positive integer fitting into a machine word is an AC^0 instruction, we expect this to be fast. In our programs, we have used the whole-number logarithm function available in our C library (`<cmath>`) which turned out to be faster than our home-made variants.

3 Space-Efficient Singly Resizable Arrays

A *singly resizable array* is a data structure that supports the grow and shrink operations at the back end plus the location of an arbitrary element, all in constant worst-case time. That is, a levelwise-allocated pile could be used for implementing a singly resizable array. In this section we describe a realization of a singly resizable array that requires only $O(\sqrt{n})$ extra space if the data structure contains n elements. The structure is similar to that presented by Brodnik et al. citeBCDMS99a,BCDMS99b, but we use a pile to explain its functioning.

Basically, our realization of a singly resizable array is nothing but a pile where each level ℓ is divided into blocks of size $2^{\lceil \ell/2 \rceil}$ and where space is allocated only for those blocks that contain elements. Therefore, we call it a *blockwise-allocated pile*. Again to avoid the allocation/deallocation problem at block boundaries, we maintain the invariant that there may exist only at most one empty block, i.e., the last empty block is released when the block prior to it gets empty. The pointers to the beginning of the blocks are stored separately in a *levelwise-allocated twin-pile*; we call this structure a twin-pile since the number of pointers at level ℓ is $2^{\lfloor \ell/2 \rfloor}$. Therefore, in a twin-pile two consecutive levels can be of the same size, but the subsequent level must be twice as large. The data structure is illustrated in Figure 2.

Since the block sizes grow geometrically, the size of the largest block is proportional to $\sqrt{n}$ if the structure stores n elements. Also, the number of blocks is proportional to $\sqrt{n}$. In the twin-pile there are at most two non-full levels. Hence, $O(\sqrt{n})$ extra space is used for pointers kept there. In the blockwise-allocated pile there are at most two non-full blocks. Hence, $O(\sqrt{n})$ extra space is reserved there for elements.

The location of an element is almost equally easy as in the levelwise-allocated pile; now only three memory accesses are necessary. Resizing is also relatively easy to implement. When the size is increased by one and the corresponding block does not exist in the blockwise-allocated pile, a new block is allocated (if

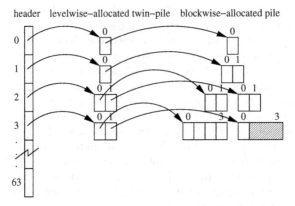

Fig. 2. A space-efficient singly resizable array storing 12 elements

there is no empty block) and a pointer to the beginning of that block is added to the end of the twin-pile as described earlier. When the size is decreased by one and the corresponding block gets empty, the space for the preceding empty block is released (if there is any); the shrinkage in the twin-pile is handled as described earlier.

One crucial property, which we use later on, is that a space-efficient singly resizable array of a given size can be constructed in reverse order and it can be used simultaneously during such a construction already after the first element is moved into the structure. Even if part of the construction is done in connection with each shrink operation, more precisely before it, the structure retains its usability during the whole construction. Furthermore, in this organization space need only be allocated for non-empty blocks in the blockwise-allocated pile and for non-empty levels in the twin-pile.

4   Space-Efficient Doubly Resizable Arrays

A *doubly resizable array* is otherwise as a singly resizable array but it can grow and shrink at both ends. We use two singly resizable arrays to emulate a doubly resizable array. We call the singly resizable arrays A and B, respectively, and the doubly resizable array emulated by these D. Assume that A and B are connected together such that A implements the changes at the front end of D and B those at the back end of D. From this the indexing of the elements is easily derived. This emulation works perfectly well unless A or B gets empty. Next we describe how this situation can be handled time- and space-efficiently.

Assume that A gets empty, and let m denote the size of B when this happens. The case where B gets empty is handled symmetrically. The basic idea is to halve B, move the first half of its elements (precisely $\lfloor m/2 \rfloor$ elements) to A, and the remaining half of its elements (precisely $\lceil m/2 \rceil$ elements) to a new B. This reorganization work is distributed for the next $\lfloor m/d \rfloor$ shrink operations to the

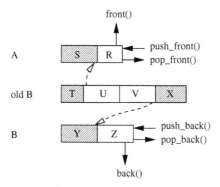

Fig. 3. Illustration of the reorganization

structure D, where $d \geq 2$ is an even integer to be determined experimentally. If $\lfloor m/d \rfloor \cdot d < m$, $\lfloor (m \bmod d)/2 \rfloor$ elements are moved to A and $\lceil (m \bmod d)/2 \rceil$ elements to B before the $\lfloor m/d \rfloor$ reorganization steps are initiated. In each of the $\lfloor m/d \rfloor$ reorganization steps, $d/2$ elements are moved from old B to A and $d/2$ elements from old B to B. The construction of A and B is done in reverse order so that they can be used immediately after they receive the first bunch of elements.

Figure 3 illustrates the reorganization. The meaning of the different zones in the figure is as follows. Zone U contains elements still to be moved from old B to A and zone S receives the elements coming from zone U. Zone R contains the elements already moved from zone T in old B to A; some of the elements moved may be erased during the reorganization and some new elements may have been inserted into zone R. Zone V contains the remaining part of old B to be moved to zone Y in B. Zone Z in B contains the elements received from zone X in old B; zone Z can receive new elements and loose old elements during the reorganization. The elements of the doubly resizable array D appear now in zones R, U, V, and Z.

If a reorganization process is active prior to the execution of a shrink operation (involving D), the following steps are carried out.

1. Take $d/2$ elements from zone U (from the end neighbouring zone T) and move them into zone S (to the end neighbouring zone R) in reverse order.
2. Take $d/2$ elements from zone V (from the end neighbouring zone X) and move them into zone Y (to the end neighbouring zone Z).

In these movements in the underlying singly resizable arrays, new blocks are allocated when necessary and old blocks are deallocated when they get empty. This way only at most a constant number of non-full blocks in the middle of the structures exists and the zones S, T, X, and Y do not consume much extra space. The same space saving is done for levels in the twin-piles.

Even if all modifying operations (involving D) done during a reorganization make A or B smaller, both of them can service at least $\lfloor m/2 \rfloor$ operations, because the work done in a single reorganization is divided for $\lfloor m/d \rfloor$ shrink operations and $d \geq 2$. Therefore, there can never be more than one reorganization process active at a time. To represent D, at most three singly resizable arrays are used. If D contains n elements, the size of A and B cannot be larger than n. Furthermore, if the size of old A or old B was m just before the reorganization started, $m \leq 2n$ at all times since the reorganization is carried out during the next $\lfloor m/d \rfloor$ shrink operations and $d \geq 2$. Hence, the number of blocks and the size of the largest block in all the three substructures is proportional to $\sqrt{n}$. That is, the bound $O(\sqrt{n})$ for the extra space needed is also valid for doubly resizable arrays.

When we want to locate the element with index k in D, we have to consider two cases. First, if no reorganization process is active, the element is searched for from A or B depending on their sizes. Let $|Z|$ denote the size of zone Z. If $k < |A|$, the element with index $|A| - k - 1$ in A is returned. If $k \geq |A|$, the element with index $k - |A|$ in B is returned. Second, if a reorganization process is active, the element is searched for from zones R, U, V, and Z depending on their sizes. If $k < |R|$, the element with index $|A| - k - 1$ in A is returned. If $|R| \leq k < |R| + |U| + |V|$, the element with index $k - |R|$ in old B is returned. If $|R| + |U| + |V| \leq k < |R| + |U| + |V| + |Z|$, the element with index $k - |R| - |U|$ in B is returned. The case where old A exists is symmetric. Clearly, the location requires only a constant number of comparisons and arithmetic operations plus an access to a singly resizable array.

5   Space-Efficient Deques

The main difference between a doubly resizable array and a deque is that a deque must also support the modifying operations. Our implementation of a space-efficient doubly resizable array can be directly used if the modifying operations simply move the elements in their respective singly resizable arrays one location backwards or forwards, depending on the modifying operation in question. This also gives the possibility to complete a reorganization process if there is one that is active. However, this will only give us linear-time modifying operations.

More efficient working is possible by implementing the blocks in the underlying singly resizable arrays circularly as proposed by Goodrich and Kloss II [ ]. If the block considered is full, a *replace* operation, which removes the first element of the block and inserts a new element at the end of the block, is easy to implement in $O(1)$ time. Only a cursor to the current first element need to be maintained; this is incremented by one (modulus the block size) and the earlier first element is replaced by the new element. A similar replacement that removes the last element of the block and adds a new element to the beginning of the block is equally easy to implement. If the block is not full, two cursors can be maintained after which replace, insert, and erase operations are all easy to accomplish.

In a space-efficient singly resizable array an insert operation inserting a new element just before the given position can be accomplished by moving the elements (after that position) in the corresponding block one position forward to make place for the new element, by adding the element that fell out of the block to the following block by executing a replace operation, and by continuing this until the last block is reached. In the last block the insertion reduces to a simple insert operation. The worst-case complexity of this operation is proportional to the number of blocks plus the size of the largest block.

An erase operation erasing the element at the given position can be carried out symmetrically. The elements after that position in the corresponding block are moved backward to fill out the hole created, the hole at the end is filled out by moving the first element of the following block here, and this filling process is repeated until the last block is reached, where the erasure reduces to a simple erase operation. Clearly, the worst-case complexity is asymptotically the same as that for the insert operation.

In a space-efficient doubly resizable array, the repeated replace strategy is applied inside A, old A/B, or B depending on which of these the modifying operation involves. Furthermore, the modifying operation involving old B (old A) should propagate to B (A). If a reorganization process is active, one step of the reorganization is executed prior to erasing an element.

Since the blocks are realized circularly, for each full block one new cursor pointing to the current first element of the circular block must be stored in the twin-piles. This will only double their size. There are a constant number of non-full blocks (at the ends of zones R, U, V, and Z); for each of these blocks one more cursor is needed for indicating the location of its last element, but these cursors require only a constant amount of extra space.

To summarize, all sequence operations run in $O(1)$ worst-case time. If the deque stores n elements, the total number of blocks in A, B, and old A/B is proportional to $\sqrt{n}$; similarly, the size of the largest block is proportional to $\sqrt{n}$. In connection with every modifying operation, in one block $O(\sqrt{n})$ elements are moved one position forwards or backwards and at most $O(\sqrt{n})$ blocks are visited. Therefore, the modifying operations run in $O(\sqrt{n})$ time.

6   Experimental Results

In this section we report the results of a series of benchmarks where the overall goal was to measure the cost of being space-efficient. This is done by comparing the efficiency of our implementation to the efficiency of SGI's implementation for the core deque operations. For reference, we have included the results for SGI's vector in our comparisons.

All benchmarks were carried out on a dual Pentium III system with 933 Mhz processors (16 KB instruction cache, 16 KB data cache and 256 KB second level cache) running RedHat Linux 6.1 (kernel version 2.2.16-3smp). The machine had 1 GB random access memory. The compiler used was gcc (version 2.95.2) and the C++ standard library shipped with this compiler included the SGI STL (version 3.3). All optimizations were enabled during compilation (using option -O6). The timings have been performed on integer containers of various sizes by using the `clock()` system call. For the constant time operations, time is reported per operation and has been calculated by measuring the total time of executing the operations and dividing this by the number of operations performed.

The first benchmarks were done to determine the best value for d. The results of these suggested that the choice of d was not very important for the performance of our data structure. For instance, doing a test using $d = 4$ in which we made just as many operations (here `pop_backs`) as there were elements to be restructured improved the time per `pop_back` by approximately 10 ns compared to $d = 2$. Increasing d to values higher than 32 did not provide any significant improvement in running times. This indicates that our data structure does not benefit noticeably from memory caching. The improvements in running times come exclusively from shortening the restructuring phase. Our choice for the value of d has thus become 4.

Next we examined the performance of the sequence operations. The best case for our data structure is when neither singly resizable array used in the emulation becomes empty, since reorganization will then never be initiated. The worst case occurs when an operation is executed during a reorganization. The following table summarizes the results for `push_back` and `pop_back` operations. Results for `push_front` and `pop_front` operations were similar for SGI's deque and our deque and are omitted; vector does not support these operations directly.

container	push_back *(ns)*	pop_back *(ns)*
`std::deque`	85	11
`std::vector`	115	2
space-efficient deque	113	35
space-efficient deque (with reorganization)	113	375

The performance of `push_back` operations for our deque is on par with that for SGI's vector, which suffers from the need to reallocate its memory from time to time. Compared to SGI's deque there is an overhead of approximately 30 percent. This overhead is expected, since there is more bookkeeping to be done for our data structure. The overhead of SGI's deque for reallocating its index block is small enough to outperform our deque.

Looking at the `pop_back` operations SGI's deque and vector are about 3 and 15 times faster than our deque when no reorganization is involved. The reason for this is that these two structures do not deallocate memory until the container itself is destroyed. For SGI's deque, the index block is never reallocated to a smaller memory block, and the same goes for the entire data block of SGI's vector. In fact, for SGI's vector the `pop_back` operation reduces to executing a single subtraction, resulting in a very low running time. This running time was verified to be equal to the running time of a single subtraction by running Bentley's instruction benchmark program, see [, Appendix 3]. When reorganization is involved, for our deque `pop_back` operations are approximately 34 times slower than SGI's deque. What is particularly expensive is that restructuring requires new memory to be allocated and elements to be moved in memory.

Accessing an element in a vector translates into an integer multiplication (or a shift), an addition, and a load or store instruction, whereas access to a deque is more complicated. Even though SGI's deque is simple, experiments reported in [ ] indicated that improving access times compared to SGI's deque is possible if we rely on shift instructions instead of division and modulus instructions. The following table gives the average access times per operation when performing repeated sequential accesses and repeated random accesses, respectively.

container	sequential access (ns)	random access (ns)
`std::deque`	117	210
`std::vector`	2	60
space-efficient deque	56	160
space-efficient deque (with reorganization)	58	162

From the results it is directly seen that, due to its contiguous allocation of memory, a vector benefits much more from caching than the other structures when the container is traversed sequentially. For SGI's deque the running time is reduced by about a factor two and for our deque the running time is reduced by about a factor three, SGI's deque being a factor two slower than our deque. As regards random access, SGI's deque is approximately 1.3 times slower than our deque, even though the work done to locate an element in our data structure comprises more instructions. To locate the data block to which an element belongs, SGI's deque needs to divide an index by the block size. The division instruction is expensive compared to other instructions (see [, Appendix 3]). In our deque, we must calculate for instance $2^{\lfloor k/2 \rfloor}$ (the number of blocks at level k), which can be expressed using left (`<<`) and right (`>>`) shifts as `1 << (k >> 1)`. Furthermore, because our data blocks are circular, we need to access elements in these blocks modulus the block size. Since our data block sizes are always powers of two, accessing element with index `i` in a block of size `b` starting at index `h` can be done by calculating `(h + i) & (b - 1)` instead of `(h + i) % b`. Modulus is just as expensive as division and avoiding it makes access in circular blocks almost as fast as access in vectors.

The improved time bounds for insert and erase operations achieved by using circular blocks are clearly evident from the benchmark results. The table below gives the results of a test inserting 1 000 elements in the middle of the container. Results for the erase operation were similar and are omitted.

container	1 000 inserts (s) initial size 10 000	1 000 inserts (s) initial size 100 000	1 000 inserts (s) initial size 1 000 000
`std::deque`	0.07	1.00	17.5
`std::vector`	0.015	0.61	12.9
space-efficient deque	0.003	0.01	0.04

With 100 000 elements in the container before the 1 000 insert operations, SGI's deque is 100 times slower than our deque, and SGI's vector is 15 times slower. The difference between $O(n)$ and $O(\sqrt{n})$ is even more clear when n is $1\,000\,000$. SGI's deque and vector are outperformed approximately by a factor 436 and factor 321, respectively.

References

1. J. BENTLEY, *Programming Pearls*, 2nd Edition, Addison-Wesley, Reading, Massachusetts (2000). 42, 48, 49
2. J. BOJESEN, Heap implementations and variations, Written Project, Department of Computing, University of Copenhagen, Copenhagen, Denmark (1998). Available at `http://www.diku.dk/research-groups/performance-engineering/resources.html`. 42
3. A. BRODNIK, S. CARLSSON, E. D. DEMAINE, J. I. MUNRO, AND R. SEDGEWICK, Resizable arrays in optimal time and space, *Proceedings of the 6th International Workshop on Algorithms and Data Structures*, Lecture Notes in Computer Science **1663**, Springer-Verlag, Berlin/Heidelberg, Germany (1999), 37–48. 41
4. A. BRODNIK, S. CARLSSON, E. D. DEMAINE, J. I. MUNRO, AND R. SEDGEWICK, Resizable arrays in optimal time and space, Technical Report CS-99-09, Department of Computer Science, University of Waterloo, Waterloo, Canada (1999). Available at `ftp://cs-archive.uwaterloo.ca/cs-archive/CS-99-09/`. 41
5. DEPARTMENT OF COMPUTING, UNIVERSITY OF COPENHAGEN, The Copenhagen STL, Website accessible at `http://cphstl.dk/` (2000–2001). 40
6. M. T. GOODRICH AND J. G. KLOSS II, Tiered vectors: Efficient dynamic arrays for rank-based sequences, *Proceedings of the 6th International Workshop on Algorithms and Data Structures*, Lecture Notes in Computer Science **1663**, Springer-Verlag, Berlin/Heidelberg, Germany (1999), 205–216. 41, 46
7. ISO (THE INTERNATIONAL ORGANIZATION FOR STANDARDIZATION) AND IEC (THE INTERNATIONAL ELECTROTECHNICAL COMMISSION), *International Standard ISO/IEC 14882: Programming Languages — C++*, Genève, Switzerland (1998). 40
8. B. B. MORTENSEN, The deque class in the Copenhagen STL: First attempt, Copenhagen STL Report 2001-4, Department of Computing, University of Copenhagen, Copenhagen, Denmark (2001). Available at `http://cphstl.dk/`. 41, 49
9. M. H. OVERMARS, *The Design of Dynamic Data Structures*, Lecture Notes in Computer Science **156**, Springer-Verlag, Berlin/Heidelberg, Germany (1983). 41
10. P. J. PLAUGER, A. A. STEPANOV, M. LEE, AND D. R. MUSSER, *The C++ Standard Template Library*, Prentice Hall PTR, Upper Saddle River, New Jersey (2001). 40
11. SILICON GRAPHICS, INC., Standard template library programmer's guide, Worldwide Web Document (1990–2001). Available at `http://www.sgi.com/tech/stl/`. 40
12. B. STROUSTRUP, *The C++ Programming Language*, 3rd Edition, Addison-Wesley Publishing Company, Reading, Massachusetts (1997). 40
13. H. SUTTER, Standard library news, part 1: Vectors and deques, *C++ Report* **11**,7 (1999). Available at `http://www.gotw.ca/publications/index.htm`. 39
14. J. W. J. WILLIAMS, Algorithm 232: Heapsort, *Communications of the ACM* **7** (1964), 347–348. 41, 42

Designing and Implementing a General Purpose Halfedge Data Structure

Hervé Brönnimann[1]

Polytechnic University
Brooklyn NY 11201, USA
hbr@photon.poly.edu

1 Introduction

Halfedge data structures (HDS) are fundamental in representing combinatorial geometric structures, useful for representing any planar structures such as plane graphs and planar maps, polyhedral surfaces and boundary representations (BREPs), two-dimensional views of a three dimensional scene, etc. Many variants have been proposed in the literature, starting with the winged-edge data structure of Baumgart[ ], the DCEL of [,], the quad-edge data structure [ ], the halfedge data structure [, , and refs. therein]. They have been proposed in various frameworks (references too many to give here):

- Plane structures: including planar maps for GIS, 2D Boolean modeling, 2D graphics, scientific computations, computer vision. The requirements on HDS are that that some edges may be infinite (e.g., Voronoi diagrams), or border edges (e.g., for bounded polygonal domains), it may include holes in the facets (planar maps), and that if so, one of the connected boundary cycle is distinguished as the outer boundary (the others are inner holes).
- Boundary representation of three-dimensional solids: including Brep representation, solid modeling, polyhedral surfaces, 3D graphics. The requirements here vary slightly: holes may still be allowed, but there is no need to distinguish an outer boundary, infinite edges are not always useful but border edges might need to be allowed.
- Planar structures encountered in higher dimensional structures: even though the data structure itself may be higher dimensional, we might want to interpret some two-dimensional substructure by using a HDS. Examples include the polygonal facets of a 3D model, or the local structure in a neighborhood of a vertex in a 3D subdivision, or the two-dimensional view of a three-dimensional scene.
- Special structures such as triangulations or simplicial complexes: in these structures, the storage is facet-based. They are usually easier to extend to higher dimensions, and a systematic presentation is given in [ ].

All the implementations we are aware of, including those surveyed above, capture a single variant of HDS, with the notable exception of the design of halfedge data structure in CGAL presented in [ ] which still limits itself to facet-based variants

G. Brodal et al. (Eds.): WAE 2001, LNCS 2141, pp. 51–66, 2001.
© Springer-Verlag Berlin Heidelberg 2001

and does not allow holes in facets, for instance, but provides some variability (forward/backward, with or without vertices, facets, and their corresponding links, see below). This design was done for visualization of 3D polyhedral surfaces, and can be reused in several situations.

Virtually everybody who has programmed a DCEL or halfedge structure knows how difficult it is to debug it and get right. Often, bugs arise when adding or removing features for reuse in a different project. Hence, the problem we deal with here is to present a design which can express as many as possible of the variants of HDS that have been proposed in the literature, in order to design a *single set* of generic algorithms that can operate on *any* of them. The goals for our halfedge data structure design are similar to those presented in [ ] :

- genericity: our specifications need to adapt to many existing structures, regardless of their internal representation. This makes it possible to express *generic algorithms* on them. Our goal is to capture all the features mentioned above. genericity implies that if features are not required, but used nevertheless (perhaps because they are required by another algorithm to be applied subsequently), the algorithm should adapt gracefully and maintain those features as well.
- power of expression: the various restrictions on HDS models in the CGAL project led to the existence of three distinct data structures, one for polyhedral surfaces [ ], one for planar maps [ ], and another one for triangulations [ ]. By contrast, we want to express all these structures using a single framework, and have a single set of algorithms to deal with these structures.
- efficiency: we do not want to sacrifice efficiency for flexibility. This entails not maintaining or interacting with unused features of a HDS, and putting minimum requirements for the algorithms manipulating an HDS. Efficiency can be achieved by using C++ templates (static binding) and compiler optimization. Also, attention needs to be paid to issues of locality of reference and memory layout.
- ease-of-use: attaching information to the vertices, edges, or facets should be easy, as well as reuse of the existing component. Also the interface needs to be uniform and easily learnable (somewhat standard). We reuse and extend the C++ STL framework of concepts and models, also know as generic programming [ ], to the case of this pointer-based data structure. See also the Boost Graph Library [ ] for similar treatment of graph algorithms.

This paper extends the excellent study presented in [ ] for the CGAL library, by providing a generic and all-purpose design for the entire family of halfedge data structures and their variants. We validate the approach in a C++ template library, the HDSTL, which is described below.

The paper is organized as follows. In the next section, we present the unified framework for working with and describing variants of HDS. Then we describe a set of generic algorithms that apply to these structures. We introduce the HDSTL, a small template library (less than 5000 lines of code) which provides a wide range of models, which we evaluate both individually and used in a practical algorithm. We conclude by evaluating how this design meets the goals expressed above.

2   Concepts

Halfedge data structures have a high degree of variability. We follow the commonality/variability analysis framework of Coplien [ ]. Briefly speaking, they may allow several representations (vertices, facet, or no none), as well as holes in facets, infinite edges (incident to a single vertex), boundary edges (incident to a single facet), etc. The representation itself may allow various access to the data structure, such as clockwise or counterclockwise traversal of a facet boundary, of a vertex cycle, access to the source or target vertex of a halfedge. Even the type of container for the components may be an array (linear storage), a list (linked storage), or other kinds of containers. But the commonality is also clear: the intent is to model 2-manifold topology, so every edge is incident to at most two facets, and in fact every halfedge is incident to at most one facet and has an opposite halfedge. Halfedges are ordered circularly along a facet boundary. Also every edge is incident to two vertices, and in fact every halfedge is incident to a unique source vertex.

In this description, a halfedge data structure is simply a structured set of pointers which satisfy some requirements. The names of those pointers and the requirements are grouped by concepts. This lets us easily describe what kind of a HDS is expected by an algorithm. The purpose of expressing concepts is to describe easily and systematically to what kind of a HDS a generic algorithm should apply. In the process of designing our concepts, we try and express invariants, and their consequences on the validity of the HDS. (The reader is referred for instance to [ ] and more recent papers by these authors for a formal approach guaranteeing consistency.) We formulate our invariants such that validity can be checked in constant time per halfedge, vertex or facet.

2.1   Halfedge Data Structure (HDS)

The simplest concept requires only halfedges. The HDS gives access to the halfedge type, the halfedge handle type, and the halfedge container. The only operations guaranteed to be supported is to create a pair of opposite halfedges, to take the opposite of a halfedge, and to access the halfedges (via the interface given by the concept of Container in the C++ STL). We say that a pointer is **valid** if it points to a halfedge that belongs to the HDS halfedge container. The only requirement is that the opposite of a halfedge h is a valid halfedge g, and that $opposite(g)$ is h itself. This is our first invariant I1.

I1. All the opposite pointers are valid, $opposite(h){\neq}h$, and $opposite(opposite(h))=h$, for any halfedge h.

In this and in the sequel, the variable h denotes a halfedge *handle*, or descriptor, not the actual halfedge element which could be a much bigger type. In

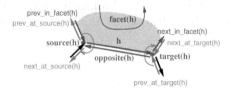

Fig. 1. The basic pointers in a HDS

the sequel, h and g denote halfedge handles, v a vertex handle, and f a facet handle.

In order to guarantee Invariant I1, halfedges are created in opposite pairs. There is thus no *new_halfedge()* function, but *new_edge()* creates two halfedges.

Remark. It usually required in the literature that the handle is a pointer, which can be more generally captured by the C++ STL concept of Trivial Iterator, meaning the following expressions are valid: **h, *h=x* (assignment, for mutable handles only), default constructor, and *h-¿m* (equivalent to *(*h).m*). Because we want to attain the maximum generality, we would also like to allow handles to be simple descriptors, like an indices in a table. This precludes the use of notation like *h-¿opposite()* as used in e.g. CGAL. We therefore assume that the access to pointers is given by the HDS itself, or using C++ notation, by *hds.opposite(h)*. The cost of generality comes at somewhat clumsier notation.

2.2   Forward/Backward/Bidirectional HDS

In order to encode the order of the halfedges on the boundary of a facet, we must have access to the halfedge immediately preceding or succeeding each halfedge h on the facet cycle. In order to encode the order of the halfedges incident to a vertex, we must have access to the halfedge immediately preceding or succeeding each halfedge h on either the source or the target vertex cycle. We have now described the pointers that are involved in our halfedge data structure: in addition to the already described *opposite()*, the pointers that link halfedges together are *next_at_source()*, *next_at_target()*, *next_in_facet()*, *prev_at_source()*, *prev_at_target()*, *prev_in_facet()*. They are shown on Figure 1.

Note that all this information need not be stored. For instance, *next_at_source(h)* is the same as *next_in_facet(opposite(h))*, while *next_at_target(h)* is the same as *opposite(next_in_facet(h))*. In practice, since we always require access to the opposite of a halfedge, it suffices to store only one of *next_at_source*, *next_at_target*, or *next_in_facet* and one has access to all three!

We call a data structure in which one has access to all three pointers *next_at_source*, *next_at_target*, and *next_in_facet*, and that satisfy the invariant I2 below, a *forward* HDS.

I2. If a HDS is forward and satisfies invariant I1, then all the pointers *next_at_source*, *next_at_target*, and *next_in_facet* are valid, and for any halfedge *h*, we have *next_at_source(h) = next_in_facet(opposite(h))*, and *next_at_target(h) = opposite(next_in_facet(h))*.

Similarly, if one of the pointers *prev_at_source*, *prev_at_target*, or *prev_in_facet* is available, then all three are, and if they satisfy the invariant I3 below, the HDS is called a **backward** HDS.

I3. If a HDS is forward and satisfies invariant I1, then all the pointers *prev_at_source*, *prev_at_target*, and *prev_in_facet* are valid, and for any halfedge *h*, we have *prev_at_source(h) = opposite(prev_in_facet(h))*, and *prev_at_target(h) = prev_in_facet(opposite(h))*.

A data structure which is both forward and backward is called **bidirectional**. We require that any HDS must provides access to either forward or backward pointers.[1] For a bidirectional HDS, we require the invariant:

I4. If a HDS is bidirectional, then *prev_at_source(next_at_source (h)) = h*, *prev_at_target (next_at_target(h)) = h*, and *prev_in_facet (next_in_facet (h)) = h*, for any halfedge *h*.

2.3   Vertex-Supporting and Facet-Supporting HDS

The basic pointers at a halfedge give an axiomatic definition of vertices and facets. A source cycle is a non-empty range of halfedges $h_0 \ldots, h_k$ such that $h_{i+1} = hds.next\_at\_source(h_i)$ for any halfedge h_i in this range and $h_k = hds.next\_at\_source(h_1)$ if the HDS is forward, or such that $h_i = hds.prev\_at\_source(h_{i+1})$ for any halfedge h_i in this range and $h_1 = hds.prev\_at\_source(h_k)$ if the HDS is backward. Likewise, a target cycle is defined similarly by using the *next_at_target* and *prev_at_target* pointers instead, and a (facet) boundary cycle by using *next_in_facet* and *prev_in_facet*. Two abstract vertices are **adjacent** if they contain at least a pair of opposite halfedges.

The important property of HDS is that each halfedge is incident to only one facet and has only two endpoints. Since the halfedge is oriented, the two vertices are therefore distinguished as the source and the target of the halfedge. We say that an HDS **supports vertices** if it provides the vertex type, the vertex handle type, access to the vertex container, as well as two pointers *source(h)* and *target(h)* for any halfedge *h*. Moreover, these pointers must satisfy the following invariants.

I5. If a HDS satisfies invariant I1 and supports vertices, then *source(g)=target(h)* and *target(g)=source(h)*, for any pair of opposite halfedges *h* and *g*.

---

[1] Note that without this requirement, our data structure would consist of unrelated pairs of opposite edges. This is useless if vertices are not supported. If they are, it might be useful to treat such a structure like a graph, without any order on the halfedges adjacent to a given vertex. Still, it would be necessary for efficient processing to have some access to all the halfedges whose source (or target) is a given vertex. This access would enumerate the halfedges in a certain order. So it appears that the requirement is fulfilled after all.

I6. If a HDS satisfies invariant I1 and supports vertices, then *source(next_in_facet(h))=target(h)* for any halfedge *h*.

Invariant I5 expresses that opposite halfedges have the same orientation, and Invariant I6 expresses that the halfedges on a boundary cycle are oriented consistently. Note that because of invariant I5, we need only store a pointer to the source or to the target of a halfedge. We may trade storage for runtime by storing both, but such a decision should be made carefully. More storage means the updates to the HDS take more time, therefore one needs to carefully evaluate whether the increased performance in following links is actually not offset by the loss of performance in setting and updating the pointers.

We express new invariants if vertices or facets are supported. Note that these invariants are checkable in linear time, and without any extra storage.

I7. If a HDS supports vertices, and satisfies invariants I1–I4, then *source(h)* = *source(g)* for any halfedges *h, g* that belong to the same source cycle, and *target(h)* = *target(g)* for any halfedges *h, g* that belong to the same target cycle

I8. If a HDS supports facets, and satisfies invariants I1–I4, then *facet(h)* = *facet(g)* for any halfedges *h, g* that belong to the same boundary cycle.

2.4   Vertex and Facet Links

Even though our HDS may support vertices or facets, we may or may not want to allocate storage from each vertex of facet to remember one (perhaps all) the incidents halfedges. We say that a vertex-supporting HDS is **source-linked** if it provides a pointer *source_cycle(v)* to a halfedge whose source is the vertex *v*, and that it is **target-linked** if it provides a pointer *target_cycle(v)* to a halfedge whose source is the vertex *v*. A facet-supporting HDS is **facet-linked** if it provides a pointer *boundary_cycle(f)* to a halfedge on the boundary of any facet (in which case it must also provide the reverse access *facet(h)* to the facet which is incident to a given halfedge *h*). It is possible to envision use of both vertex- and facet-linked HDS, and non-linked HDS. The following invariants guarantee the validity of the HDS.

I9. If a HDS supports vertices, is source-linked, and satisfies Invariants I1–I7, then *source(source_cycle(v))=v* for every vertex *v*.

I10. If a HDS supports vertices, is target-linked, and satisfies Invariants I1–I7, then *target(target_cycle(v))=v* for every vertex *v*.

I11. If a HDS supports facets, is facet-linked, and satisfies Invariants I1–I6 and I8, then *facet(boundary_cycle(f))=f* for every facet *f*.

2.5   HDS with Holes in Facets and Singular Vertices

An HDS may or may not allow facets to have **holes**. Not having holes means that each facet boundary consists of a single cycle; it also means that there is a one-to-one correspondence between facets and abstract facets. In a HDS supporting holes in facets, each facet is required to give access to a hole container.[2] This

---

[2] The container concept is defined in the C++ STL.

Fig. 2. An illustration of (a) facets with holes, (b) outer boundary, and (c) singular vertices

container may be global to the HDS, or contained in the facet itself. Each element of that container need only point to a single halfedge.

In a facet with holes, one of the cycles may be distinguished and called the *outer boundary*; the other holes are the *inner holes*. This is only meaningful for plane structure (see Figure 2(b)), where the outer boundary is distinguished by its orientation which differs from that of the inner holes. In Figure 2(a), for instance, the outer boundary is defined if we know that the facet is embedded in a plane, but there is no non-geometric way to define the outer boundary of the grayed facet: a topological inversion may bring the inside boundary cycle outside. (Note that when the incident facet is by convention to the left of the halfedge, the outer boundary is oriented counterclockwise, while the inner hole boundaries are oriented clockwise.)

In a connected HDS without holes, it is possible to reconstruct the facets without the facet links. In general, for an HDS with holes but which is not facet linked, it is impossible to reconstruct the facets as the information concerning which cycles belong to the same facet is lost, let alone which is the outer boundary. Therefore, if the HDS supports facets with holes, we require that it be facet-linked, and that it also provide the outer boundary and an iterator over the holes of any facet.

An HDS may also allow singular vertices. A vertex is *singular* if its corresponding set of adjacent halfedges consists of several cycles; it is the exact dual notion of hole in facet and has the same requirements concerning cycle container.

In a singular vertex, one of the cycles may be distinguished and called the *outer cycle*, and the other cycles are called *inner cycles*. They are depicted in Figure 2(c). Unlike the outer facet boundary, the possibility of inner cycles seems to only make sense for dual structures (see section on duality below) and may not be really useful, although it perhaps may be useful in triangulating 3D polyhedra (e.g. the algorithm of Chazelle and Palios keeps an outer boundary and may have inner vertex cycles when removing a dome). As with holes, if an HDS supports singular vertices, we require that it be vertex-linked and that it provide the outer cycle and an iterator over the inner cycles of any vertex.

2.6   Hole Links, Vertex Cycle Links, Infinite and Border Edges, Duality

By analogy with vertex and facet links, we call the HDS **source-cycle-linked** if it provides a pointer *source_cycle(h)* for each halfedge *h*, **target-cycle-linked** if it provides a pointer *target_cycle(h)* for each halfedge *h*, and **hole-linked** if it provides a pointer *boundary_cycle(h)* for each halfedge *h*.

An edge whose facet pointer has a singular value is called a **border** halfedge. Likewise, a halfedge whose source or target pointer has a singular value is called an **infinite** halfedge.

For lack of space, we cannot elaborate on these notions. Suffice it to note that with all these definitions, our set of concepts is closed under duality transformations which transform vertices into facets and vice-versa.

2.7   Memory Layout

In addition to the previous variabilities which have to do with the functionality of the data structure, and the model it represents, we offer some support for memory layouts. The only requirement we have made so far on the layout concerns the availability of halfedge, vertex and facet containers, as well as vertex cycle and hole containers if those are supported.

A **mutable** structure allows modification of its internal pointers, via functions such as *set_opposite(h,g)*, etc. These functions need only be supported if the corresponding pointer is accessed: a forward HDS is only required to provide *set_next_in_facet()*, *set_next_at_source()*, and *set_next_at_target()*. The reason we require separate functions for read or write access, is that a pointer may be accessed even though it cannot be set (if this pointer is non-relocatable, see next section).

There is considerable freedom in the implementation of an HDS. For instance, because of invariants I2 and I3, it is desirable to have pairs of opposite halfedges close in memory, so a possible way to do this as suggested by Kettner [ ] is to store them contiguously. In this case the opposite pointer may be implicit and its storage may be reclaimed. The same trick could be used for any of the other pointers (such as *next_in_facet*, etc.).

Another example is the CGAL triangulation, which can be captured in our framework as a HDS which is normalized by facets thus saving the storage for *all* bidirectional pointers. We call this layout a **triangulated HDS**. There are thus six pointers only per triangular facet (three opposite and three source vertex pointers), which matches the CGAL triangulation data structure [ ]. A restricted set of algorithms needs to be applied (*new_edge* replaced by *new_triangle*). This is unavoidable since a triangulated HDS cannot express all the possible states of a HDS. Note that some triangulation algorithms can maintain and work with a triangulated HDS (iterative and sweep algorithms) but others cannot because of their need of general HDS as intermediate representation (e.g., divide-and-conquer, see below).

Because of this freedom, we need to introduce one more concept: an HDS is *halfedge-relocatable* with respect to a given pointer if any two halfedge locations can be exchanged in the container, and the pointers to these halfedges updated, without affecting the validity of the HDS. An HDS which actually stores all its pointers is halfedge-relocatable, while the HDS given as example above, which store pairs of opposite halfedges contiguously, is not. Similar definitions can be made for *vertex-* and *facet-relocatable* HDS. These concepts are important in discussing normalization algorithms (see Section 3.3).

A halfedge-relocatable HDS provides a function *halfedge_relocate(h,g)* to relocate a halfedge pointed to by h to a position pointed to by g (whatever was at that position is then lost). It also provides a member function *halfedge_swap(h,g)* for swapping two halfedges in the containers without modifying the pointers in the HDS.

3   Generic Algorithms

The purpose of enunciating a collection of concepts is to describe precisely how the algorithms interact with a data structure. In the C++ STL, this interaction is achieved very elegantly through iterators: containers provide iterators, whose semantics can be modified by adapters, and algorithms operate on a range of iterators. In the HDSTL, the interaction takes place through a set of functors (such as *opposite, next_in_facet,* etc.), and whose arguments and return types are handles. Using this framework, we can express operators and algorithms on a HDS that specify exactly what the (compile-time) requirements are, and what the semantics of the algorithm are.

3.1   Elementary Algorithms

In order to write generic algorithms, we need elementary manipulations of pointers. These are complicated by the fact that these pointers may or may not be supported, even the corresponding types may not exist. So we need "smart" functions which either provide the pointer or a default-constructed value if the pointer is not supported. The elementary functions come in several sets: the *get_...., set_...., copy_...., compare_...., create/destroy_....* functions.[3] In addition, in a valid HDS, the *next_...* and *prev_....* pointers may be computed by a reverse traversal of a (vertex or boundary) cycle, stopping before the cycle repeats. These are the *find_....* functions. For convenience, we provide *stitch_....* functions which set pairs of reverse pointers.

In [ ], these functions exist and are encapsulated in a so-called decorator. We choose to provide them in the global scope since they are the primary access to the HDS. Their first argument is always an object of type HDS.

---

[3] The *compare_....* functions are especially useful to write pre- and post-conditions for the high-level operations.

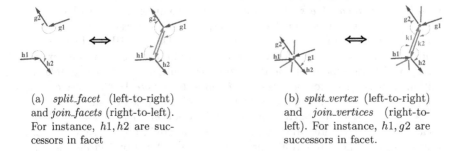

(a) *split_facet* (left-to-right) and *join_facets* (right-to-left). For instance, $h1, h2$ are successors in facet

(b) *split_vertex* (left-to-right) and *join_vertices* (right-to-left). For instance, $h1, g2$ are successors in facet.

Fig. 3. the meaning of the arguments *h1,h2,g1,g2* for the Euler operators

3.2 Euler and oOther Combinatorial Operators

In addition to the elementary functions, most algorithms use the same set of high-level operations, called Euler operations since they preserve the Euler number of the structure $(v - e + f - i - 2(c + g)$, where v, e, f, i , c, are the numbers of vertices, edges, facets, inner holes, and connected components, and g is the genus of the map). The basic operation used by these operators is splice which breaks a vertex cycle by inserting a range of edges. Using *splice*, we provide an implementation of dual operator *join_facets* and *join_vertices* (delete a pair of opposite halfedges and merge both adjacent boundary or vertex cycles), and of their reverse *split_facet* and *split_vertices*. By using the *get_...*, *set_...* and *stitch_...* functions we can write operators that work seamlessly on any kind of HDS (forward, backward, bidirectional, etc.).

As mentioned above, this variability does not come without consequences. For one, the Euler operators must be passed a few more parameters. Most operators take at least four arguments, such as *split_facet (h1,h2,g1,g2)* or *join_facets(k1,k2,h1,h2,g1,g2)* with the meaning depicted in Figure 3. For convenience, there are also functions such as *split_facet_after (h1,g1)* if the HDS is forward (the parameters *h2* and *g2* can be deduced), or *join_facets(k1)* if the HDS is bidirectional.

Some preconditions are an integral part of the specifications: for instance, *join_facets* has the precondition that either facet links are not supported, or that the boundary cycles differ for both halfedges. For HDS with holes, this operation is provided and called *split_boundary* and its effect is similar to *join_facets,* except that for HDS with holes, it records the portion containing *g1* as an inner hole of the common adjacent facet. (This part is still under integration in the HDSTL.)

3.3 Memory Layout Reorganization

An important part of designing a data structure is the memory layout. Most implementations impose their layout, but our design goal flexibility implies that we give some control to the user as well. We differentiate with the static design (fixed at compile time, Section 2.7) and the dynamic memory layout reorganization which is the topic of this paragraph.

We say that a HDS is *normalized by facets* if the halfedge belonging to boundary cycles are stored contiguously, in their order along the boundary. This means that the halfedge iterator provided by the HDS will enumerate each facet boundary in turn. In case facets are supported, moreover, *facet normalization* means that the corresponding facets are also stored in the same order. A HDS is *normalized by (source or target) vertices* if a similar condition is satisfied for (source or target) vertex cycles, with similar definition of *vertex normalization,* and *normalized by opposite* if halfedges are stored next to their opposite. These algorithms can only be provided if the HDS is halfedge-relocatable. Note that in general, these normalizations are mutually exclusive. Furthermore, they do not get rid of the storage for the pointers (except in the case where the length of the cycle is fixed, such as for opposite pointers, or in triangulations, in which case some implicit scheme can save storage, as explained in Section 2.7). Normalization algorithms can be implemented in linear time.

4 Experimental Evaluation

4.1 Comparison of Various Models

Here we compare various models of HDS provided in the HDSTL. The models can be *compact* (normalized by opposite, hence no storage for opposite pointer), *indices* (handles are integers instead of pointers; dereferencing is more costly, but copying does not necessitate pointer translation), or pointer-based by default. The functionality can be *minimal* (no vertices or facets), *graphic_fw* (vertices, no facets, and forward-only), *graphic* (vertices but no facets), *default* (vertices, facets, but forward-only), and *maximal* (vertices, facets, bidirectional). Moreover, we have the possibility of storing the halfedge incidences at source (*next_at_source*) or at target (*next_at_target*) instead of in facet (*next_in_facet*), as well as choosing between storing the source or target vertex. We do this in order to measure the differences in run time. We measure no difference with storing incidences at source or at target.

The last three lines illustrate as well other HDS for which we could write adaptors. We have not mentioned runtime ratios, because it is not clear how to provide a fair comparison. The LEDA graph [ ] stores vertices, edges, and halfedges in a doubly linked list, and stores four edge pointers per node and per edge, as well as degree per node and other extra information. It is therefore a very powerful, but also very fat structure, and it offers no flexibility. The CGAL models of HDS [ ] however, are comparable in storage and functionality, although less flexible. They require for instance facet-based pointer storage (*next* and *prev_in_facet*). We can reasonably estimate that their performance would match the corresponding models in the HDSTL.

The results are shown in Table 1. We measure the running time ratios by effecting a million pairs *split_facet/join_facets* in a small (tetrahedron) HDS.

Measurements with million pairs *split_vertices*/*join_vertices* are slightly slower, but in similar ratio. There is therefore no effect due to the (cache and main) memory hierarchy. The running times can vary between a relative 0.6 (facet-based minimal using vector storage) to 2.2 (source-based maximal using list storage). The index-based structure is also slower, but not by much, and it has the double advantage that copying can be done directly without handle translation, and that text-mode debugging is facilitated because handles are more readable.

In conclusion, the model of HDS can affect the performance of a routine by a factor of more than 4. In particular, providing less functionality is faster because fewer pointers have to be set up. Of course, if those pointers are used heavily in the later course of a program, it might be better to store them than recompute them. We also notice that the runtime cost of more functionality is not prohibitive, but care has to be taken that the storage does not exceed the main storage available, otherwise disk swapping can occur. When dealing with large data, it is best to hand pick the least memory-hungry model of HDS that fits the requirements of the program. Also, for high-performance computing applications, its might be a good strategy to "downgrade" the HDS (allow bidirectional storage but declare the HDS as forward-only) and in a later pass set the extra pointers.

4.2   Delaunay Triangulation

We programmed the divide-and-conquer Delaunay triangulation [,], with Dwyer's trick for efficiency (first partition in vertical slabs of $\sqrt{n} \log n$ points each, recursively process these slabs by splitting with horizontal lines, then merge these slabs two by two). There does not seem to be a way to do with only forward HDS: when merging two triangulations, we need a clockwise traversal for one and a counter-clockwise for the other; moreover, simply storing the convex hull in a bidirectional list will not do, as we may need to remove those edges and access the edges inside during the merging process. We need vertices (to contain a point) and bidirectional access to the HDS. This algorithm only uses orientation and in-circle geometric tests. We used a custom very small geometry kernel with simple types (array of two floating point *double* for points).

For this reason, we only tried the divide-and-conquer algorithm with three variants of HDS: the graphic HDS (vertices and edges, bidirectional, with vertex links) the compact graphic HDS (graphic without vertex links, with opposite halfedges stored contiguously), and the maximal HDS (graphic with facets and facet links as well). Used by the algorithm are the vertex information, opposite and bidirectional pointers in facets and at vertices. The algorithm does not use but otherwise correctly sets all the other pointers of the HDS.

The results of our experiments are shown in Table 2. We are not primarily interested in the speed of our implementation, although we note that on a Pentium III 500Mhz with sufficient memory (256MB), triangulating a million points takes about 11 seconds for the base time, which is very competitive. More interestingly, however, we compare again the relative speed of the algorithms with our different models of HDS, and we record both the running time

Table 1. A synoptic view of the different models of HDS : (a) name, (b) features [V=vertex,F=facet,L=link,FW=forward,BI=bidirectional, if=in_facet,as=at_source], (c) number of pointers required by the structure, (d) runtime ratios for pair split_facet/join_facets

HDS	features	pointers	runtime ratio
default	V+L,FWif,F+L	v+4h+f	0.95
	id., FWas	v+4h+f	1.0
	id., list storage	3v+6h+3f	2.0
indices	default+indices	v+4h+f	1.25
	id., FWas	v+h+f	1.25
compact_fw	FW,compact	h	0.6
compact	BI,compact	2h	0.65
minimal	FWif	2h	0.6
	id. but FWas	2h	0.65
graphic_fw	V+L,FWif	v+3h	0.7
	id., FWas	v+3h	0.75
graphic	V+L,BIif	v+4h	0.7
	id., BIas	v+4h	0.75
maximal	V+L, BIif, F+L	2v+5h+2f	1
	id., BIas	2v+5h+2f	1.1
	id., list storage	4v+7h+4f	2.15
LEDA_graph	V+L, BIas	$\geq 4(v+h)$	N/A
CGAL_minimal	FWif	2h	N/A
CGAL_maximal	V+L, BIif, F+L	2v+5h+2f	N/A

and the memory requirements. The results in Table 2 suggest that the model of the HDS has an incidence on the behavior of the algorithm, although more in the memory requirement than in the running time (there is at most 15% of variation in the running time). Notice how the compact data structure uses less memory and also provokes fewer page faults (measured by the Unix *time* command). Also interesting is the difference between the compact and non-compact versions of the graphic HDS with and without Dwyer's strategy (10% vs. only 2%). These experiments seem to encourage the use of compact data structures (when appropriate).

These experiments also show how the functional requirements of an algorithm limit the choices of HDS that can be used in them. We could not test the graphic-forward HDS with this algorithm. When provided, however, the comparison is fundamentally fair since it is the *same* algorithm which is used with all the different structures .

Table 2. A comparison of different HDS used with divide-and-conquer Delaunay triangulation without (first three lines) or with (last three lines) Dwyer's trick (and 2.10^5 points): (a) running times and (b) ratios, (c) memory requirements, (d) paging behavior

HDS	time	ratio	memory	page faults
graphic_cp	6.635	1.11	22.3MB	24464
graphic	5.95	1	27.9MB	29928
maximal	6.9	1.15	35.6MB	36180
graphic_cp	3.22	1.03	22.3MB	24464
graphic	3.125	1	27.9MB	29928
maximal	3.57	1.145	35.6MB	36180

5   Conclusion

We have provided a framework for expressing several (indeed, many) variants of the halfedge data structure (HDS), including geometric graph, winged-edge, Facet/Edge and dual Vertex/Edge structures. Our approach is to specify a set of requirements (refining each other) in order to express the functional power of a certain HDS. This approach is a hallmark of generic programming, already used in the C++ STL, but also notably in the Boost Graph Library and in CGAL [12]. Regarding the latter reference, we have enriched the palette of models that can be expressed in the framework. For instance, we can express holes in facets or non-manifold vertices. Also we do not require any geometry, thus allowing the possibility to use the HDSTL in representing purely combinatorial information (for instance, in representing hierarchical maps in GIS).

As a proof of concept, we offer an implementation in the form of a C++ template library, the HDSTL (halfedge data structure template library), which consists of only 4,500 lines of (admittedly dense) code, but can express a wide range of structures summarized on Table 1. We provide a set of generic algorithms (including elementary manipulations, Euler operators, and memory layout reorganization) to support those structures.

Regarding our goals, flexibility is more than achieved, trading storage cost for functionality: We have provided a rich variety of models in our framework, each which its own strength and functionality. We also have shown that it is possible to provide a set of generic algorithms which can operate on every single one of them. This flexibility is important when the HDS is expressed as a view from within another structure (e.g. solid boundary, lower-dimensional section or projection, dual view) for which we may not control the representation. It is of course always possible to set up a HDS by duplicating the storage, but problems arise from maintaining consistency. Instead, our flexibility allows to access a view in a coherent manner by reusing the HDSTL algorithms, but on the original storage viewed differently.

Ease of use comes from the high-quality documentation provided with the HDSTL, but there are issues with error messages when trying to use unsupported features, or when debugging. This can be improved using recent work in concept checking [ ].

Efficiency is acquired by using C++ templates (static binding, which allows further optimizations) instead of virtual functions (dynamic binding, as provided in Java). The algorithms we provide are reasonably efficient (roughly fifteen seconds for Delaunay triangulation of a million points on a Pentium 500Mhz with sufficient main memory).

Acknowledgments

Thanks to Lutz Kettner for many discussions about his design, to Jack Snoeyink for his helping hand in the early debugging process, and to the people in the CGAL project, notably Sylvain Pion and Monique Teillaud, for past discussions and inspiration.

References

1. M. H. Austern. *Generic Programming and the STL*. Professional computing series. Addison-Wesley, 1999. 52
2. B. G. Baumgart. A polyhedron representation for computer vision. In *Proc. AFIPS Natl. Comput. Conf.*, volume 44, pages 589–596. AFIPS Press, Arlington, Va., 1975. 51
3. J.-D. Boissonnat, O. Devillers, M. Teillaud, and M. Yvinec. Triangulations in CGAL. In *Proc. 16th Annu. ACM Sympos. Comput. Geom.*, pages 11–18, 2000. 52, 58
4. E. Brisson. Representing geometric structures in d dimensions: Topology and order. *Discrete Comput. Geom.*, 9:387–426, 1993. 51
5. C. Burnikel. Delaunay graphs by divide and conquer. Research Report MPI-I-98-1-027, Max-Planck-Institut für Informatik, Saarbrücken, 1998. 24 pages. 62
6. D. Cazier and J.-F. Dufourd. Rewriting-based derivation of efficient algorithms to build planar subdivisions. In Werner Purgathofer, editor, *12th Spring Conference on Computer Graphics*, pages 45–54, 1996. 53
7. *The CGAL Reference Manual*, 1999. Release 2.0. 61
8. J. O. Coplien. *Multi-Paradigm Design for C++*. Addison-Wesley, 1999. 53
9. M. de Berg, M. van Kreveld, M. Overmars, and O. Schwarzkopf. *Computational Geometry: Algorithms and Applications*. Springer-Verlag, Berlin, 1997. 51
10. E. Flato, D. Halperin, I. Hanniel, and O. Nechushtan. The design and implementation of planar maps in CGAL. In *Abstracts 15th European Workshop Comput. Geom.*, pages 169–172. INRIA Sophia-Antipolis, 1999. 52
11. L. J. Guibas and J. Stolfi. Primitives for the manipulation of general subdivisions and the computation of Voronoi diagrams. *ACM Trans. Graph.*, 4(2):74–123, April 1985. 51
12. L. Kettner. Using generic programming for designing a data structure for polyhedral surfaces. *Comput. Geom. Theory Appl.*, 13:65–90, 1999. 51, 52, 58, 59, 64
13. L.-Q. Lee, J. G. Siek, and A. Lumsdaine. The generic graph component library. In *Proceedings OOPSLA'99*, 1999. 52

14. K. Mehlhorn and S. Näher. *LEDA: A Platform for Combinatorial and Geometric Computing.* Cambridge University Press, Cambridge, UK, 1999. 61
15. D. E. Muller and F. P. Preparata. Finding the intersection of two convex polyhedra. *Theoret. Comput. Sci.*, 7:217–236, 1978. 51
16. F. P. Preparata and M. I. Shamos. *Computational Geometry: An Introduction.* Springer-Verlag, 3rd edition, October 1990. 62
17. J. Siek and A. Lumsdaine. Concept checking: Binding parametric polymorphism in C++. In *First Workshop on C++ Template Programming, Erfurt, Germany,* October 10 2000. 65
18. K. Weiler. *Topological Structures for Geometric Modeling.* PhD thesis, Rensselaer Polytechnic Institute, Troy, NY, August 1986. 51

Optimised Predecessor Data Structures for Internal Memory*

Naila Rahman[1], Richard Cole[2], and Rajeev Raman[3]

[1] Department of Computer Science, King's College London
`naila@dcs.kcl.ac.uk`
[2] Computer Science Department, Courant Institute, New York University
`cole@cs.nyu.edu`
[3] Department of Maths and Computer Science, University of Leicester
`r.raman@mcs.le.ac.uk`

Abstract. We demonstrate the importance of reducing misses in the *translation-lookaside buffer (TLB)* for obtaining good performance on modern computer architectures. We focus on data structures for the dynamic predecessor problem: to maintain a set S of keys from a totally ordered universe under insertions, deletions and predecessor queries. We give two general techniques for simultaneously reducing cache and TLB misses: simulating 3-level hierarchical memory algorithms and cache-oblivious algorithms. We give preliminary experimental results which demonstrate that data structures based on these ideas outperform data structures which are based on minimising cache misses alone, namely B-tree variants.

1 Introduction

Most algorithms are analysed on the random-access machine (RAM) model of computation, using some variety of the unit-cost criterion. This postulates that accessing a location in memory costs the same as a built-in arithmetic operation, such as adding two word-sized operands. Over the last 20 years or so CPU clock rates have grown explosively, but the speeds of main memory have not increased anywhere near as rapidly. Nowadays accessing a main memory location may be over a hundred times slower than performing an operation on data held in the CPU's registers. In addition, accessing (off-chip) main memory dissipates much more energy than accessing on-chip locations [ ], which is especially undesirable for portable and mobile devices where battery life is an important consideration.

In order to reduce main memory accesses, modern computers have multiple levels of *cache* between CPU and memory. A cache is a fast associative memory, often located on the same chip as the CPU, which holds the values of some main memory locations. If the CPU requests the contents of a main memory location, and the value of that location is held in some level of cache, the CPU's request is

* Research supported in part by EPSRC grant GR/L92150 (Rahman, Raman), NSF grant CCR-98-00085 (Cole) and UISTRF project 2001.04/IT (Raman).

G. Brodal et al. (Eds.): WAE 2001, LNCS 2141, pp. 67–78, 2001.
© Springer-Verlag Berlin Heidelberg 2001

answered by the cache itself (a cache *hit*); otherwise it is answered by consulting the main memory (a cache *miss*). A cache hit has small penalty (1-3 cycles is typical) but a cache miss is very expensive. To amortise the cost of a main memory access, an entire *block* of consecutive main memory locations which contains the location accessed is brought into cache on a miss. Thus, a program that exhibits good *locality*, i.e. one that accesses memory locations close to recently accessed locations, will incur fewer cache misses and will consequently run faster and consume less power. Much recent work has focussed on analysing cache misses in algorithms [1,15,17,16,18,13,6]. Asymptotically, one can minimise cache misses by emulating optimal 2-level external-memory algorithms [8], for which there is a large literature [21].

There is another source of main memory accesses which is frequently overlooked, namely misses in the *translation-lookaside buffer (TLB)*. The TLB is used to support *virtual memory* in multi-processing operating systems. Virtual memory means that the memory addresses accessed by a process refer to its own unique logical address space, and it is considered "essential to current computer systems" [ ]. To implement virtual memory, most operating systems partition main memory and the logical address space of each process into contiguous fixed-size *pages*, and store (recently-accessed) logical pages of each active process in main memory at a time [1]. This means that every time a process accesses a memory location, the reference to the corresponding logical page must be translated to a physical page reference. Unfortunately, the necessary information is held in the *page table*, a data structure in main memory! To avoid looking up the page table on each memory access, translations of some recently-accessed logical pages are held in the TLB, which is a fast associative memory. If a memory access results in a TLB hit, there is no delay, but a TLB miss can be much more expensive than a cache miss.

In most computer systems, a memory access can result in a TLB miss alone, a cache miss alone, neither, or both: algorithms which make few cache misses can nevertheless make many TLB misses. Cache and TLB sizes are usually small compared to the size of main memory. Typically, cache sizes vary from 256KB to 8MB, and TLBs hold between 64 and 128 entries. Hence *simultaneous* locality at the cache and page level is needed even for internal-memory computation.

In [ ] a model incorporating both cache and TLB was introduced, and it was shown that radix sort algorithms that optimised both cache and TLB misses were significantly faster than algorithms that optimised cache misses alone. Indeed, the latter made many more TLB misses than cache misses. In [ ] it was also shown that by emulating optimal 2-level external-memory algorithms one ensures that the numbers of TLB and cache misses are of the same order of magnitude, and the number cache misses is optimal. This minimises the *sum* of cache and TLB misses (asymptotically). However, this is not enough. In [ ] it is argued that due to the different characteristics of a cache miss and a TLB miss—in particular, a TLB miss can be more expensive than a cache miss—these

---

[1] In case a logical page accessed is not present in main memory at all, it is brought in from disk. The time for this is not counted in the CPU times reported.

should be counted separately. Even if cache and TLB misses cost about the same, optimising cache and TLB misses individually will usually reduce memory accesses by nearly a half. For example, sorting n keys requires $\Theta((n/B)\log_?(n))$ cache misses, where B is the cache block size, and $\Theta((n/P)\log_?(n))$ TLB misses, where P is the page size. Ignoring the bases of the logarithms for the moment, the fact that $P \gg B$ (our machine's values of $P = 2048$ and $B = 16$ are representative) means that the latter should be much smaller than the former. We argue that the same holds for searching, the problem that this paper focusses on.

In this paper we suggest two techniques for simultaneously minimising cache and TLB misses. The first is to simulate algorithms for a 3-level hierarchical memory model (3-HMM). The second is the use of *cache-oblivious* algorithms [ ]. Cache-oblivious algorithms achieve optimal performance in a two-level memory but without knowledge of parameters such as the cache block size and the number of cache blocks. As a consequence, they minimise transfers across each level of a multi-level hierarchical memory.

The problem we consider is the dynamic predecessor problem, which involves maintaining a set S of pairs $\langle x, i \rangle$ where x is a key drawn from a totally ordered universe and i is some satellite data. Without loss of generality we assume that i is just a pointer. We assume that keys are fixed-size, and for now we assume that each pair in S has a unique key. The operations permitted on the set include insertion and deletion of $\langle$key, satellite data$\rangle$ pairs, and predecessor searching, which takes a key q and returns a pair $\langle x, i \rangle \in S$ such that x is the largest key in S satisfying $x \leq q$. For this problem the B-tree data structure (DS) makes an optimal $O(\log_B n)$ cache misses and executes $O(\log n)$ instructions[2] for all operations, where $|S| = n$ [,]. However, a B-tree also makes $\Omega(\log_B n)$ TLB misses, which is not optimal. In this paper we consider the practical performance of three implementations.

Firstly, we consider the B*-tree [ ], a B-tree variant that gives a shallower tree than the original. We observe that by *paginating* a B*-tree we get much better performance. Pagination consists of placing as much of the top of the tree into a single page of memory as possible, and then recursing on the roots of the subtrees that remain when the top is removed. As the only significant difference between the paged and original B*-trees is the TLB behaviour, we conclude that TLB misses do significantly change the running-time performance of B*-trees. Unfortunately, we do not yet have a satisfactory practical approach for maintaining paged B-trees under insertions and deletions.

The second is an implementation of a DS which uses the standard idea for 3-HMM. At the top level the DS is a B-tree where each node fits in a page. Inside a node, the 'splitter' keys are stored in a B-tree as well, each of whose nodes fits in a cache block. This DS makes $O(\log_B n)$ cache misses and $O(\log_P n)$ TLB misses for all operations (the update cost is amortised).

Finally, we consider a cache-oblivious tree. The tree of Bender et al. [ ] makes the optimal $O(\log_B n)$ cache misses and $O(\log_P n)$ TLB misses. However it ap-

---

[2] Logs are to base 2 unless explicitly noted otherwise.

pears to be too complex to get good practical performance. Our implementation is based on a simpler cache-oblivious DS due to Bender et al [ ]. This data stucture makes $O(\log_B n + \log \log n)$ cache misses and $O(\log_P n + \log \log n)$ TLB misses.

Our preliminary experimental figures suggest that the relative performance is generally as follows:

$$B^* < \text{Cache-oblivious} < \text{3-HMM} < \text{Paged B}^*.$$

It is noteworthy that a non-trivial DS such as the cache-oblivious tree can outperform a simple, proven *and* cache-optimised DS such as the B^* tree.

2   Models Used

The basic model we use was introduced in [ ] and augmented earlier cache models [,] with a TLB. The model views main memory as being partitioned into equal-sized and aligned *blocks* of B memory words each. The cache consists of S *sets* each consisting of $a \geq 1$ *lines*. a is called the *associativity* of the cache, and typical values are $a = 1, 2$ or 4. Each line can store one memory block, and the i-th memory block can be stored in any of the lines in set $(i \bmod S)$. We let $C = aS$. If the program accesses a location in block i, and block i is not in the cache, then one of the blocks in the set $(i \bmod S)$ is *evicted* or copied back to main memory, and block i is copied into the set in its place. We assume that blocks are evicted from a set on a *least recently used (LRU)* basis. For the TLB, we consider main memory as being partitioned into equal-sized and aligned *pages* of P memory words each. A TLB holds address translation information for at most T pages. If the program accesses a memory location which belongs to (logical) page i, and the TLB holds the translation for page i, the contents of the TLB do not change. If the TLB does not hold the translation for page i, the translation for page i is brought into the TLB, and the translation for some other page is removed on a LRU basis. Our results on searching do not depend strongly on the associativity or the cache/TLB replacement policy.

We now give some assumptions about this model, which are justified in [ ]:
(A1) TLB and cache misses happen independently, i.e., a memory access may result in a cache miss, a TLB miss, neither, or both;
(A2) B, C, P and T are all powers of two, and
(A3) i. $B \ll P$;   ii. $T \ll C$;   iii. $BC \leq PT$;   iv. $T \leq P, B \leq C$.

The performance measures are the number of instructions, the number of TLB misses and the number of cache misses, all counted separately.

Our experiments are performed on a Sun UltraSparc-II, which has a word size of 4 bytes and a block size of 64 bytes, giving $B = 16$. Its L2 (main) cache, which has an associativity $a = 1$ (i.e. it is direct-mapped), holds $C = 8192$ blocks and its TLB holds $T = 64$ entries [ ]. Finally, the page size is 8192 bytes, giving $P = 2048$ and also (coincidentally) giving $BC = PT$.

3   General Techniques

An Emulation Theorem Let $\mathbf{C}(a, B, C, P, T)$ be the model described in Section 2, where a is the associativity of the cache, B is the block size, C is the capacity of the cache in blocks, P is the page size and T is the number of TLB entries. Let $\mathbf{I}(B, M, B', M')$ denote the following model. There is a fast main memory, which is organised as M/B blocks $m_1, \ldots, m_{M/B}$ of B words each, a secondary memory which is organised as M'/B' blocks $d_1, d_2, \ldots, d_{M'/B'}$ of B' words each, and an tertiary memory, which is organised as an unbounded number of blocks $t_1, t_2, \ldots$ of B' words each as well. We assume that $B \leq B'$ and that B and B' are both powers of two. An algorithm in this model performs computations on the data in main memory, or else it performs an *I/O step*, which either copies a block of size B to or from main and secondary memory, or copies a block of size B' to or from secondary and tertiary memory.

Theorem 1. *An algorithm A in the $\mathbf{I}(B, BC/4, P, PT/4)$ model which performs I_1 I/Os between main and secondary memory and I_2 I/Os between secondary and tertiary memory, and t operations on data in main memory can be converted into an equivalent one A' in the $\mathbf{C}(a, B, C, P, T)$ model which performs at most $O(t + I_1 \cdot B + I_2)$ operations, $O(I_1 + I_2)$ cache misses and $O(I_2)$ TLB misses.*

Proof: The emulation assumes that the contents of tertiary memory are stored in an array $Tert$, each of whose elements can store B' words. The emulation maintains an array $Main$ of size $C/4 + 2$, each entry of which can hold B words. The first $C/4$ locations of $Main$ emulate the main memory of A, with m_i corresponding to $Main[i]$, for $i = 1, \ldots, C/4$. The last two locations of $Main$ are buffers that are used for copying data to avoid conflict misses [ ]. For convenience, we assume that the arrays Mem and $Tert$ are aligned on cache block and page boundaries respectively.

We now assume an TLB with $T/2+2$ entries, but with a replacement strategy that we specify. By definition, the optimal (offline) replacement strategy will make no more misses than our strategy. Then we note that an LRU TLB with T entries never makes more than a constant factor more misses than an optimal (offline) TLB with $T/2+2$ entries [ ], and so the LRU TLB with T entries makes no more than a constant factor more misses than our replacement strategy. Our TLB replacement strategy will set aside $T/4$ TLB entries $\delta_1, \ldots, \delta_{T/4}$ to hold translations for the blocks which are currently in secondary memory. In particular, if (tertiary) block t_i is held in (secondary) block d_j then δ_j holds the translation for the page containing $Tert[i]$. Also, our TLB replacement strategy will never replace the translations for the pages containing $Main$. Since $Main$ occupies at most $(C/4 + 2)B \leq (T/4 + 2)P$ words, only $T/4 + 2$ TLB entries are required for this.

A read from t_i to d_j (tertiary to secondary) is simulated by "touching" the page containing $Tert[i]$ (e.g. reading the first location of $Tert[i]$). Our TLB replacement algorithm brings in the translation for $Tert[i]$ and stores it in TLB buffer δ_j. This causes one TLB miss and $O(1)$ cache misses. A read or write from d_i to m_j (primary to secondary) is done as in [ ]. Since all translations are held in TLB, the only cost of this step is $O(1)$ amortised cache misses and $O(B)$ instructions. Finally, emulating the operation of the algorithm on data in main memory incurs no cache or TLB misses.This proves the theorem.   □

Cache-oblivious Data Structures As noted in [,] a cache-oblivious algorithm gives optimal performance across each level of a multi-level hierarchy. By our emulation theorem, an optimal cache-oblivious algorithm makes an optimal number of cache and TLB misses (in general, the emulation is needed as some cache-oblivious algorithms are analysed on an 'ideal cache' model, which assumes an omniscient replacement policy).

4   Data Structures

We now describe the three DSs in more detail, starting with the cache oblivious search tree, then the paged B*-tree and then the 3-HMM search tree. Below, the term 'I/O' refers to both TLB and cache misses.

Cache-oblivious Search Tree Prokop [ ] showed how to cache-obliviously lay out a complete binary search tree T with n keys so that on a fast memory with an unknown block size of L, searches take $O(\log_L n)$ I/Os. If h is the height of T, Prokop divides T at $\lfloor h/2 \rfloor$, which separates T into the top subtree T_0 of height $\lfloor h/2 \rfloor$ and $k \leq 2^{\lfloor h/2 \rfloor}$ subtrees $T_1, \ldots, T_k$ of height $\lceil h/2 \rceil$. T is laid out recursively in memory as T_0 followed by $T_1, \ldots, T_k$, as shown in Figure 1. This DS is static and requires $O(n)$ time and $O(n/L)$ I/Os to construct. Bender et al. [ ] note that Prokop's tree can be simply and efficiently dynamised by using *exponential* search trees [ ]. Our algorithm is closely related to that of Bender et al [ ], which we briefly review now.

We first set out some terms we will use in the discussion below. An exponential search tree has non-leaf nodes of varying and sometimes quite large (non-constant) degree. A non-leaf node with k children stores $k - 1$ sorted keys that guide a search into the correct child: we will organise these $k - 1$ keys as a Prokop tree. In what follows, we refer to the exponential search tree as the *external tree*, and its nodes as *external nodes*. We refer to any of the Prokop trees in an external node as *internal trees* and to their nodes as *internal nodes*.

Roughly speaking, the root of an external tree with n leaves contains $\Theta(\sqrt{n})$ keys which partition the keys into sets of size $\Theta(\sqrt{n})$. [3] After this, we recurse on each set. In contrast to [ ], we end the recursion when we reach sets

---

[3] In reality a "bottom-up" definition is used, whereby all leaves of the external tree are at the same depth and an external node with height i has an ideal number of children that grows doubly exponentially with i. The root of the external tree may have much fewer children than its ideal.

of size $\Theta(\log n)$. These are then placed in *external leaf* nodes, which are represented as arrays. Although the keys inside external nodes are organised as Prokop trees, we do not require external nodes to have any particular memory layout with respect to each other (see Figure 2). As shown in [ ], the number of I/Os during searches, excluding searching in the external leaves, is at most $O(\log_L n + \log \log n)$. Searching in the external leaves requires $O((\log n)/L)$ I/Os, which is negligible. As shown in [ ], the DS (excluding the external leaves) can be updated within the same I/O bounds as a search. Insertions of new keys take place at external leaves and if the size of an external leaf exceeds twice its ideal size of $\Theta(\log n)$ then it is split, inserting a new key into its parent external node. To maintain invariants as n changes, the DS is rebuilt when n squares.

Implementation details. We have implemented two versions of our cache oblivious DS. In both versions the keys in an external leaf node are stored in a sorted array, and the description below is about the different structures of non-leaf external and internal nodes.

In the first version of the DS internal nodes are simply the keys of the external node or they are pointers to external children node. In order to reach the left or right child of an internal non-leaf node we calculate its address in the known memory layout. An external node with n' keys has an internal tree of height $\lceil \log n' \rceil + 1$, the first $\lceil \log n' \rceil$ levels hold the keys and the last level holds pointers to external child nodes. Each internal tree is a complete binary search tree and if $n' + 1$ is not a power of two then infinite valued keys are added to the first $\lceil \log n' \rceil$ levels of the internal tree, and NULL pointers are added to the last level. All $\lceil \log n' \rceil + 1$ levels of the internal tree are laid out recursively in a manner similar to Prokop's scheme, thus reducing the number of I/Os in moving from the last level of keys in the internal tree to the pointers to the external children nodes.

In the second version of the DS each internal node consists of a key and left and right pointers. An external node with n' keys has an internal tree of height $\lceil \log n' \rceil$. At height $h < \lceil \log n' \rceil$ the internal node pointers point to internal child nodes. At the last level of the tree internal node pointers point to external children nodes. As above, each internal node is a complete binary tree. This implementation has the disadvantage of requiring extra memory but has the advantage that it avoids the somewhat computationally expensive task of finding a child in the internal tree.

Since the insertion time in an external leaf is anyway $\Theta(\log n)$ (we insert a new key into an array), we reduce memory usage by ensuring that a external leaf with k keys is stored in a block of memory sufficient for $k + O(1)$ keys. Increasing this to $O(k)$ would improve insertion times, at the cost of increased memory. A final note is that the emulation of Section 3 is not required in this case to achieve optimal cache and TLB performance.

Paged B*-Trees B*-trees are a variant of B-trees where the nodes remain at least 66% full by sharing with a sibling node the keys in a node which has exceeded its maximum size [ ]. By packing more keys in a node, B*-trees are

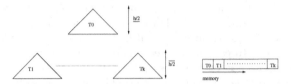

Fig. 1. The recursive memory layout of a complete binary search tree according to Prokop's scheme

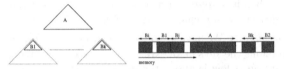

Fig. 2. Memory layout of the cache oblivious search tree with an external root node A of degree $\sqrt{n}$. The subtrees of size $\Theta(\sqrt{n})$ rooted at A are shown in outline, and their roots are $B_1, \ldots, B_k$. The memory locations used by the subtrees under $B_1, \ldots, B_k$ are not shown

shallower than B-trees. Assuming the maximum branching factor of a node is selected such that a node fits inside a cache block, a heuristic argument suggests that for sufficiently large n, the number of cache and TLB misses for the B-tree or B*-tree on random data should be about $\log_B n - \log_B C + O(1)$ and $\log_B n - \log_B T + O(1)$ respectively. This is because one may expect roughly the top $\log_B C$ levels of the B-tree to be in cache, and address translations for all nodes in the top $\log_B T$ levels to be in the TLB. A reasonable assumption for internal memory is that $C \geq \sqrt{n}$ and that $\log_B T$ term may be ignored. This gives us a rough total of $\log_B n$ TLB misses and at most $0.5 \log_B n$ cache misses. In other words, TLB misses dominate.

A *paged* B- or B*-tree uses the following memory layout of the nodes in the tree. Starting from the root and in a breadth-first manner as many nodes as possible are allocated from the same memory page. The sub-trees that remain ouside the page are recursively organised in a similar manner (see Fig 3). The process of laying out a B-tree in this way is called *pagination*. The number of cache misses in a paged B-tree are roughly the same as an ordinary B-tree, but the number of TLB misses is sharply reduced from $\log_B n$ to about $\log_P n$. With our values of B and P the TLB misses are reduced by about two-thirds, and the overall number of misses is reduced by about a half. Unfortunately, we can only support update operations on a paged B-tree if the B-tree is weight-balanced [1,22], but such trees seem to have poorer constant factors than ordinary B-trees.

Optimal 3-HMM Trees We now describe the implementation of a DS that is optimised for the 3-HMM. In principle, the idea is quite simple: we make a B-tree with branching factor $\Theta(P)$ and store the splitter keys inside a node in

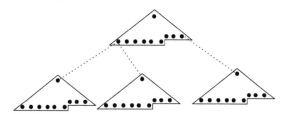

Fig. 3. A paged B-tree. The nodes in the top part of the tree are in a memory page. The top parts of sub-trees that remain outside the page with the root are recursively placed on a memory page

a B-tree with branching factor $\Theta(B)$. However, the need to make these nodes dynamic causes some problems. A B-tree with branching factor in the range $[d + 1, 2d]$ and which has a total of m nodes can store (roughly) between md and $m \cdot (2d - 1)$ keys. These correspond to trees with branching factor $d + 1$ and $2d$ at all nodes, respectively. If we let m be the number of inner B-tree nodes which fit in a page, the *maximum* branching factor of the outer B-tree cannot exceed md, otherwise the number of nodes needed to store this tree may not fit in a page. Since each inner B-tree node takes at least $4d$ words ($2d$ keys and $2d$ pointers), we have that $m \le P/(4d)$ and thus the maximum branching factor of the outer B-tree is at most $P/4$.

To make the tree more bushy, we simply rebuild the entire inner B-tree whenever we want to add an (outer) child to a node, at a cost of $\Theta(P)$ operations. When rebuilding, we make the inner B-trees have the maximum possible branching factor, thus increasing the maximum branching factor of the outer B-tree to about $P/2$. To compensate for the increased update time, we apply this algorithm only to the non-leaf nodes in the outer B-tree, and use standard B-tree update algorithms at the leaves. This again reduces the maximum branching factor at the leaves to $P/4$. At the leaves, however, the problem has a more serious aspect: the fullness of the leaves is roughly half of what it would be compared to a comparable standard B-tree. This roughly doubles the memory usage relative to a comparable standard B-tree.

Some of these constants may be reduced by using B*-trees in place of B-trees, but the problem remains significant. A number of techniques may be used to overcome this, including using an additional layer of buckets, overflow pages for the leaves etc, but these all have some associated disadvantages. At the moment we have not fully investigated all possible approaches to this problem.

5   Experimental Results

We have implemented B*-trees, paged B*-trees, optimal 3-HMM trees and our cache oblivious search trees with and without internal pointers. Our DSs were coded in C++ and all code was compiled using gcc 2.8.1 with optimisation level 6. Our experiments were performed on a Sun UltraSparc-II with 2×300 Mhz processors and 1GB main memory. This machine has 64 TLBs, 8KB pages, a 16KB L1 cache, and a 512KB L2 cache; both caches are direct-mapped. The L1 cache miss penalty is about 3 cycles on this machine, and the L2 cache and TLB miss penalties are both about 30 cycles.

In each experiment, a DS was built on pairs of random 4-byte keys and 4-byte satellite data presented in random order. It should be noted that random data tends to degrade cache performance, so these tests are hard rather than easy. The B*-tree and cache-oblivious trees were built by repeated insertions, as essentially were the 3-HMM trees.

B*-trees and paged B*-trees were tested with branching factors of 7 and 8 keys per node (allowing 8 keys and 8 pointers to fit in one cache block of 64 bytes). Paged B*-trees were paginated for 8KB pages. For each DS on n keys, we performed searches for $2n$ fresh keys drawn from the same distribution (thus "successful" searches are rare). These were either $2n$ independently generated keys, or $n/512$ independent keys, each of which was searched for 1024 times in succession. The latter minimises the effect on cache misses and thus estimates the average computation cost. For each algorithm, we have measured the average search time per query. Insertion costs are not reported. These are preliminary results and at each data point we report the average search time per query for 10 experiments.

Each DS was tested with $n = 2^{18}, 2^{20}, 2^{22}$ and 2^{23} uniform integer keys in the range $[0, 2^{31})$. At each data point we report the average search time ("abs") and below it the average search time relative to the fastest average search time at that data point ("rel"). Figure 4 shows results for the random queries and Figure 5 shows results for the repeated random queries.

In Fig 4 we see that paged B*-trees are by far the fastest for random queries. Random queries on B*-trees, tuned just for the cache, take between 43% and almost 70% more time than on paged B*-trees with a branching factor of 7. Optimal 3-HMM trees perform quite well, being at-most 14% slower than paged B*-trees. Of the DSs suited for both the cache and TLB, the cache oblivious search trees are the slowest. On the other hand, the 3-HMM trees are carefully tuned to our machine, whereas the cache-oblivious trees have no machine-specific parameters.

As can be seen from Figure 5, the better-performing DSs do not benefit from code tweaks to minimise operation costs. In fact, as they are a bit more complex, they actually have generally higher operation costs than B-trees, especially the cache-oblivious DS with implicit pointers. Thus one would expect even better relative performance for the search trees suited to both the cache and TLB versus B*-trees on machines with higher miss penalties—the cache and TLB miss penalties are quite low on our machines.

It is also instructive to compare these times with classical search trees such as Red-Black trees. E.g., in an earlier study with the same machine/OS/compiler combination, one of the authors reported a time of 9.01 μs/search for $n = 2^{22}$ using LEDA 3.7 Red-Black trees [, Fig 1(a)]. This is over twice as slow as B*-

trees and over thrice as slow as paged B*-trees. LEDA (a, b) trees are a shade slower than our B*-trees.

	Time per-search (μs) on UltraSparc-II 2 $\times$ 300 Mhz						
n	B* (BF=7)	B* (BF=8)	Paged B* (BF=7)	Paged B* (BF=8)	Optimal 3-HMM	Cache Obl. (int. ptrs)	Cache Obl. (no int. ptrs)
2^{18} abs	2.384	2.216	1.541	1.572	1.655	1.875	2.109
2^{18} rel	**1.547**	**1.438**	**1.000**	**1.020**	**1.074**	**1.217**	**1.369**
2^{20} abs	3.329	3.054	1.979	2.398	2.210	2.563	2.770
2^{20} rel	**1.682**	**1.543**	**1.000**	**1.212**	**1.117**	**1.295**	**1.400**
2^{22} abs	4.113	3.797	2.487	2.913	2.843	3.299	3.369
2^{22} rel	**1.654**	**1.527**	**1.000**	**1.171**	**1.143**	**1.326**	**1.355**
2^{23} abs	4.529	4.156	2.691	3.127	3.054	3.635	3.658
2^{23} rel	**1.683**	**1.544**	**1.000**	**1.162**	**1.135**	**1.351**	**1.359**

Fig. 4. Query times for $2n$ independent random queries on n keys in DS

	Computation time per-search (μs) on UltraSparc-II 2 $\times$ 300 Mhz machine						
n	B* (BF=7)	B* (BF=8)	Paged B* (BF=7)	Paged B* (BF=8)	Optimal 3-HMM	Cache Obl. (int. ptrs)	Cache Obl. (no int. ptrs)
2^{18} abs	0.793	0.751	0.736	0.744	0.904	0.885	1.262
2^{18} rel	**1.077**	**1.020**	**1.000**	**1.011**	**1.228**	**1.202**	**1.715**
2^{20} abs	0.891	0.868	0.867	0.862	1.024	0.941	1.315
2^{20} rel	**1.034**	**1.007**	**1.006**	**1.000**	**1.188**	**1.092**	**1.526**
2^{22} abs	0.976	0.955	0.962	0.947	1.129	1.170	1.399
2^{22} rel	**1.031**	**1.008**	**1.016**	**1.000**	**1.192**	**1.235**	**1.477**
2^{23} abs	0.982	0.979	0.975	0.966	1.151	1.244	1.401
2^{23} rel	**1.017**	**1.013**	**1.009**	**1.000**	**1.192**	**1.288**	**1.450**

Fig. 5. Query times for $n/512$ independent random queries repeated 1024 times each, on n keys in DS

6   Conclusions and Future Work

Our preliminary experimental results show that optimising for three-level memories gives large performance gains in internal memory computations as well. In particular we have shown that cache-oblivious data structures may have significant practical importance. Although the trends in our preliminary experiments are clear, these need to be rigorously established with a larger suite of experiments, including cache and TLB simulations. We would also like to test performance of these structures on secondary memory.

References

1. L. Arge, J. S. Vitter. Optimal Dynamic Interval Management in External Memory (extended abstract). *FOCS 1996*, pp. 560-569. 74
2. A. Andersson. Faster Deterministic Sorting and Searching in Linear Space. In *Proc. 37th IEEE FOCS*, pp. 135–141, 1996. 72
3. Bender, M., Cole, R. and Raman. R. Exponential trees for cache-oblivious algorithms. In preparation, 2001. 70, 72, 73
4. Bender, M., Demaine, E. and Farach-Colton, M. Cache-oblivous B-trees. In *Proc. 41st IEEE FOCS*, pp. 399–409, 2000. 69
5. Comer, D. The Ubiquitous B-Tree. *ACM Comput. Surv.* **11** (1979), p.121. 69, 73
6. Frigo, M., Leiserson, C. E., Prokop, H., and Ramachandran, S. Cache-oblivious algorithms. In *Proc. 40th IEEE FOCS*, pp. 285–298, 1999. 68, 69, 72
7. Furber, S. B. *Arm System-On-Chip Architecture* Addison-Wesley Professional, 2nd ed., 2000. 67
8. Hennessy, J. L. and Patterson, D. A. *Computer Architecture: A Quantitative Approach* (Second ed.). Morgan Kaufmann, 1996. 68
9. D. E. Knuth. *The Art of Computer Programming. Volume 3: Sorting and Searching, 3rd ed.* Addison-Wesley, 1997. 69
10. Ladner, R. E., Fix, J. D., and LaMarca, A. Cache performance analysis of traversals and random accesses. In *Proc. 10th ACM-SIAM SODA* (1999), pp. 613–622.
11. LaMarca, A. and Ladner, R. E. The influence of caches on the performance of sorting. *J. Algorithms* **31**, 66–104, 1999. 68, 70
12. Korda, M. and Raman, R. An experimental evaluation of hybrid data structures for searching. In *Proc. 3rd WAE*, LNCS 1668, pp. 213–227, 1999. 76
13. Mehlhorn, K. and Sanders, P. Accessing multiple sequences through set-associative cache, 2000. Prel. vers. *Proc. 26th ICALP*, LNCS 1555, 1999. 68, 70
14. H. Prokop. Cache-oblivious algorithms. MS Thesis, MIT, 1999. 72
15. Rahman, N. and Raman, R. Analysing cache effects in distribution sorting. *ACM J. Exper. Algorithmics*, WAE '99 special issue, to appear. Prel. vers. in *Proc. 3rd WAE*, LNCS 1668, pp. 184–198, 1999. 68
16. Rahman, N. and Raman, R. Analysing the cache behaviour of non-uniform distribution sorting algorithms. In *Proc. 8th ESA*, LNCS 1879, pp. 380–391, 2000. 68
17. Rahman, N. and Raman, R. Adapting radix sort to the memory hierarchy. TR 00-02, King's College London, 2000, `http://www.dcs.kcl.ac.uk/technical-reports/2000.html` Prel. vers. in *Proc. ALENEX 2000*. 68, 70
18. Sen, S. and Chatterjee, S. Towards a theory of cache-efficient algorithms (extended abstract). In *Proc. 11th ACM-SIAM SODA* (2000), pp. 829–838. 68, 71, 72
19. Sleator, D. D. and Tarjan, R. E. Amortized efficiency of list update and paging rules. *Communications of the ACM* **28**, 202–208, 1995. 71
20. Sun Microsystem. UltraSPARC User's Manual. Sun Microsystems, 1997. 70
21. Vitter, J. S. External memory algorithms and data structures: Dealing with MASSIVE data. To appear in *ACM Computing Surveys*, 2000. 68
22. D. E. Willard. Reduced memory space for multi-dimensional search trees. In *Proc. STACS '85*, pages 363–374, 1985. 74

An Adaptable and Extensible Geometry Kernel

Susan Hert[1], Michael Hoffmann[2], Lutz Kettner[3], Sylvain Pion[4], and
Michael Seel[1]

[1] Max-Planck-Institut für Informatik
Stuhlsatzenhausweg 85, 66123 Saarbrücken, Germany
{hert,seel}@mpi-sb.mpg.de
[2] Institute for Theoretical Computer Science, ETH Zurich
CH-8092 Zurich, Switzerland
hoffmann@inf.ethz.ch
[3] University of North Carolina at Chapel Hill, USA
kettner@cs.unc.edu
[4] INRIA, Sophia Antipolis - France
Sylvain.Pion@sophia.inria.fr

Abstract. Geometric algorithms are based on geometric objects such
as points, lines and circles. The term *Kernel* refers to a collection of representations for constant-size geometric objects and operations on these
representations. This paper describes how such a geometry kernel can be
designed and implemented in C++, having special emphasis on adaptability, extensibility and efficiency. We achieve these goals following the
generic programming paradigm and using templates as our tools. These
ideas are realized and tested in CGAL [ ], the Computational Geometry
Algorithms Library.

Keywords: Computational geometry, library design, generic programming.

1 Introduction

Geometric algorithms that manipulate constant-size objects such as circles, lines,
and points are usually described independent of any particular representation of
the objects. It is assumed that these objects have certain operations defined on
them and that simple predicates exist that can be used, for example, to compare
two objects or to determine their relative position. Algorithms are described in
this way because all representations are equally valid as far as the correctness
of an algorithm is concerned. Also, algorithms can be more concisely described
and are more easily seen as being applicable in many settings when they are
described in this more generic way.

We illustrate here that one can achieve the same advantages when implementing algorithms by encapsulating the representation of objects and the operations
and predicates for the objects into a geometry kernel. Algorithms interact with
geometric objects only through the operations defined in the kernel. This means
that the same implementation of an algorithm can be used with many different

G. Brodal et al. (Eds.): WAE 2001, LNCS 2141, pp. 79–90, 2001.
© Springer-Verlag Berlin Heidelberg 2001

representations for the geometric objects. Thus, the representation can be chosen to be the one most applicable (*e.g.*, the most robust or most efficient) for a particular setting.

Regardless of the representation chosen by a particular kernel, it cannot hope to satisfy the needs of every application. For example, for some applications one may wish to maintain additional information with each point during the execution of an algorithm or one may wish to apply a two-dimensional algorithm to a set of coplanar points in three dimensions. Both of these things are easily accomplished if the kernel is implemented to allow types and operations to be redefined, that is, if the kernel is easily adaptable. It is equally important that a kernel be extensible since some applications may require not simply modifications of existing objects and operations but addition of new ones.

Although adaptability and extensibility are important and worthwhile goals to strive for, one has to keep in mind that the elements of the kernel form the very basic and fundamental building blocks of a geometric algorithm built on top. Hence, we are not willing to accept *any* loss in efficiency on the kernel level. Indeed, using template programming techniques one can achieve genericity without sacrifying runtime-performance by resolving the arising overhead during compile-time.

After discussing previous work on the design of geometry kernels (Section 2), we give a general description of our new kernel concept (Section 3). We then describe how this concept can be realized in an adaptable and extensible way under the generic programming paradigm [21,22] (Sections 4 through 7). Section 8 illustrates the use of such a kernel and shows how the benefits described above are realized. Finally, we describe the models of this type of kernel that are provided in CGAL (Section 9).

As our implementation is in C++ [ ], we assume the reader is familiar with this language; see [2,17,26] for good introductions.

2   Motivation and Previous Work

Over the past 10 years, a number of geometry libraries have been developed, each with its own notion of a geometry kernel. The C++ libraries PLAGEO and SPAGEO [ ] provide kernels for 2- and 3-dimensional objects using floating point arithmetic, a class hierarchy, and a common base class. The C++ library LEDA [ ] provides in its geometry part two kernels, one using exact rational arithmetic and the other floating point arithmetic. The Java library GEOMLIB [ ] provides a kernel built in a hierarchical manner and designed around Java interfaces. None has addressed the questions of easily exchangeable and adaptable kernels.

Flexibility is one of the cornerstones of CGAL [ ], the *Computational Geometry Algorithms Library*, which is being developed in a common project of several universities and research institutes in Europe and Israel. The recent overview [ ] gives an extensive account of functionality, design, and implementation techniques in the library. Generic programming is one of the tools used to achieve this flexibility [6,21,22].

In the original design of the geometry kernel of CGAL [ ], there was a representation class which encapsulates how geometric objects are represented. These representation classes could be easily exchanged or extended, and they provided some limited adaptability. However, the design did not allow the representation classes to also include geometric operations. This extension was seen as desirable after the introduction of *geometric traits classes* into the library, which separate the combinatorial part of an algorithm or data structure from the underlying geometry. The term traits class was originally introduced by Myers [ ]; we use it here to refer to a class that aggregates (geometric) types and operations. By supplying different traits classes, the same algorithm can be applied to different kinds of objects. The fact that the existing CGAL kernel did not present its functionality in a way that was immediately accessible for the use in traits classes was one motivation for this work. Factoring out common requirements from the traits classes of different algorithms into the kernel is very helpful in maintaining uniform interfaces across a library and maximizing code reuse.

While the new design described here is even more flexible and more powerful than the old design, it maintains backwards compatibility. The kernel concept now includes easily exchangeable functors in addition to the geometric types; the ideas of traits classes and kernel representations have been unified. The implementation is accomplished by using a template programming idiom similar to the Barton-Nackman technique [,] that uses a derived class as a template argument for a base class template. A similar idiom has been used in CGAL to solve cyclic template dependencies in the halfedge data structure and polyhedral surface design [ ].

3   The Kernel Concept and Architecture

A geometry kernel K consists of types used to represent geometric objects and operations on these types. Since different kernels will have different notions of what basic types and operations are required, we do not concern ourselves here with listing the particular objects and operations to be included in the kernel. Rather, we describe the kernel concept in terms of the interface it provides for each object and operation.

Depending on one's perspective, the expected interface to these types and operations will look somewhat different. From the point of view of an imperative-style programmer, it is natural that the types appear as stand-alone classes and the operations as global functions or member functions of these classes.

```
K::Point_2   p(0,1), q(1,-4);
K::Line_2    line(p, q);
if (less_xy_2(p, q)) { ... }
```

However, from the point of view of someone implementing algorithms in a generic way, it is most natural, indeed most useful, if types and operations are both provided by the kernel. This encapsulation allows both types and operations to be adapted and exchanged in the same manner.

```
K k;
K::Construct_line_2  c_line   = k.construct_line_2_object();
K::Less_xy_2         less_xy  = k.less_xy_2_object();
K::Point_2           p(0,1);
K::Point_2           q(1,-4);
K::Line_2            line = c_line(p, q);
if (less_xy(p, q)) { ... }
```

The concept of a kernel we introduce here includes both of these perspectives. That is, each operation is represented both as a type, an instance of which can be used like a function, and as a global function or a member function of one of the object classes. The techniques described in the following three sections allow both interfaces to coexist peacefully under one roof with a minimal maintenance overhead, and thus lead to a kernel that presents a good face to everyone.

Our kernel is constructed from three layers, illustrated in Figure 1. The bottom layer consists of basic numeric primitives such as the computation of matrix determinants and the construction of line equations from point coordinates. These numeric primitives are used in the geometric primitives that constitute the second layer of our structure. The top layer then aggregates and assimilates the geometric primitives. The scope of our kernel concept is representation-independent affine geometry. Thus the concept includes, for example, the construction of a point as the intersection of two lines but not its construction from x and y coordinates.

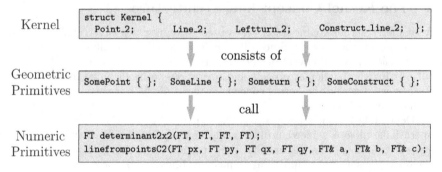

Fig. 1. The kernel architecture

4   An Adaptable Kernel

We present our techniques using a simplified example kernel. Consider the types Point_2 and Line_2 representing two-dimensional points and lines, respectively, and an operation Construct_line_2 that constructs a Line_2 from two Point_2 arguments. In general, one probably needs more operations and possibly more types in order to be able to do something useful, but for the sake of simplicity we will stay with these four items for the time being.

A first question might be: Construct_line_2 has to construct a Line_2 from two Point_2s; hence it has to know something about both types. How does it get to know them? Since we are talking about adaptability, just hard-wiring the corresponding classnames is not what we would like to do.

A natural solution is to parameterize the geometric classes with the kernel. As soon as a class knows the kernel it resides in, it also knows all related classes and operations. A straightforward way to implement this parameterization is to supply the kernel as a template argument to the geometric classes.

```
template < class K > struct MyPoint { ... };
template < class K > struct MyLine { ... };
template < class K > struct MyConstruct { ... };
```

Then our kernel class looks as follows.

```
struct Kernel {
  typedef MyPoint< Kernel >      Point_2;
  typedef MyLine< Kernel >       Line_2;
  typedef MyConstruct< Kernel >  Construct_line_2;
};
```

At first, it might look a bit awkward; inserting a class into its own components seems to create cyclic references. Indeed, the technique we present here is about properly resolving such cyclic dependencies.

Let us come back to the main theme: adaptability. It should be easy to extend or adapt this kernel and indeed, all that needs to be done is to derive a new class from `Kernel` where new types can be added and existing ones can be exchanged.

```
struct New_kernel : public Kernel {
  typedef NewPoint< New_kernel >   Point_2;
  typedef MyLeftTurn< New_kernel > Left_turn_2;
};
```

The class `Point_2` is overwritten with a different type and a new operation `Left_turn_2` is defined. But there is a problem: the inherited class `MyConstruct` is still parameterized with `Kernel`, hence it operates on the old point class `MyPoint`. What can be done to tell `MyConstruct` that it should now consider itself being part of `New_kernel`?

An obvious solution would be to redefine `Construct_line_2` in `New_kernel` appropriately, *i.e.* by parameterizing `MyConstruct` with `New_kernel`. This is fine in our example where it amounts to just one more typedef, but considering a real kernel with dozens of types and hundreds of operations, it would be really tedious to have to repeat all these definitions.

Fortunately, there is a way out. If `Kernel` is meant as a base for building custom kernel classes, it is not wise to fix the parameterization (this process is called *instantiation*) of `MyPoint<>`, `MyLine<>` and `MyConstruct<>` at that point to `Kernel`, as this might not be the kernel in which these classes finally end up. We rather would like to defer the instantiation, until it is clear what the actual kernel will be. This can be done by introducing a class `Kernel_base` that serves as an "instantiation-engine." Actual kernel classes derive from `Kernel_base` and finally start the instantiation by injecting themselves into the base class.

```
template < class K >
struct Kernel_base {
   typedef MyPoint< K >      Point_2;
   typedef MyLine< K >       Line_2;
   typedef MyConstruct< K >  Construct_line_2;
};
struct Kernel : public Kernel_base< Kernel > {};
```

In order to be able to extend **New_kernel** in the same way as **Kernel**, we can defer instantiation once again. The construction is depicted in Figure 2.

```
template < class K >
struct New_kernel_base : public Kernel_base< K > {
   typedef NewPoint< K >    Point_2;
   typedef MyLeftTurn< K > Left_turn_2;
};
struct New_kernel : public New_kernel_base< New_kernel > {};
```

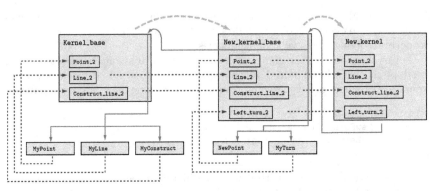

Fig. 2. Deferring instantiation. Boxes stand for classes, thick dashed arrows denote derivation, solid arrows show (template) parameters, and thin dotted arrows have to be read as "defines" (typedef or inheritance)

Thus we achieve our easily extensible and adaptable kernel through the use of the kernel as a parameter at two different levels. The geometric object classes are parameterized with the kernel such that they have a way of discovering the types of the other objects and operations. And the kernel itself is derived from a base class that is parameterized with the kernel, which assures that any modified types or operations live in the same kernel as the ones inherited from the base class and there is no problem in using the two together. Note again that this design does not create any runtime overhead, as the lookup of the correct types and operations is to be handled during compile time.

5   Functors

The question still remains how we provide the actual functionality needed by the classes and functions that interact through the kernel. There are a number of ways functions can be provided in a way that assures adaptability of the kernel. However, efficiency is also very important since many of the predicates and constructions are small, consisting of only a few lines of code. Therefore, the calling overhead has to be minimal.

The classic C-style approach would be to use *pointers to functions*, where adaptability is provided by the ability to change the pointer. *Virtual functions* are the Java-style means of achieving adaptability. In both cases though, there is an additional calling indirection involved; moreover, many compiler optimisations are not possible through virtual functions [ ], as the actual types are not known at compile time. This overhead is considerable in our context [ ].

The solution we propose is more in line with the standard C++ library [ ], where many algorithms are parameterized with so-called *function objects*, or *functors*. A functor is an abstraction of a function; that is, it is anything that behaves as a function and thus, can be used as a function. It is something you can call by using parentheses and passing arguments [ ]. Obviously, a function is a functor; but also objects of a class-type that define an appropriate operator() can be functors. There are some advantages that make this abstraction worthwhile.

Efficiency If the complete class definition is known at compile time, the operator() can be inlined. Handing this function object as a parameter to some other functor is like handing over a piece of code that can be inlined and optimized to the compiler's taste.

Functors Have State Functors also prove to be more flexible; a functor of class-type can carry local data. For example, the functor Less_int from above can easily be modified to count the number of comparisons done. Other examples of state in a functor are the binders binder1st and binder2nd in the STL. They use a local variable to store the value to which one of the two arguments of a binary adaptable functor gets bound.

Allowing local data for a functor adds a complication to the kernel. Clearly, a generic algorithm has to be oblivious to whether a functor carries local state or not. Hence, the algorithm cannot instantiate the functor itself. But we can assume that the kernel knows how to create functors. So we add access member functions to the kernel that allow a generic algorithm to obtain an object for a functor.

6   An Imperative Interface

Someone used to imperative-style programming might expect an interface based on member functions and global functions operating on the geometric classes rather than having to deal with functors and kernel objects. Due to the flexibility in our design, we can easily provide such an interface on top of the kernel with little overhead. For example, there is a global function

```
bool left_turn_2(Point_2 p, Point_2 q, Point_2 r) { ... }
```

which calls the corresponding functor Left_turn_2 in the kernel where the points p, q and r originate from. Some care has to be taken, to define these functions in a proper way such that they operate on the kernel in a truly generic manner.

Similarly, one might also want to add some functionality to the geometric types; for example a constructor to the line class `MyLine` that takes two point arguments. Again it is important that MyLine does not make assumptions about the point type, but uses only the operations provided by the kernel. This way, the geometric types remain nicely separated, as their – sometimes close – relationships are encapsulated into appropriate operations.

7   A Function Toolbox

Our kernel concept nicely separates the representation of geometric objects from the operations on these objects. But when implementing a specific operation such as `Left_turn_2`, the representation of the corresponding point type `Point_2` will inevitably come into play; in the end, the predicate is evaluated using arithmetic operations on some number type. The nontrivial[1] algebraic computations needed in predicates and constructions are encapsulated in the bottom layer of our kernel architecture (Figure 1), the *number-type-based function toolbox*, which we describe in this section.

A *number type* refers to a numerical type that we use to store coordinates and to calculate results. Given that the coordinates we start with are rational numbers, it suffices to compute within the domain of rational numbers. For certain operations we will go beyond rational arithmetic and require roots. However, since the majority of our kernel requires only rational arithmetic we focus on this aspect here. Depending on the calculations required for certain operations, we distinguish between different concepts of number types that are taken from algebra. A *ring* supports addition, subtraction and multiplication. A *Euclidean ring* supports the three ring operations and an integral division with remainder, which allows the calculation of greatest common divisors used, *e.g.*, to cancel common factors in fractions. In contrast, a *field* type supports exact division instead of integral division.

Many of the operations in our kernel boil down to determinant evaluations, *e.g.*, sidedness tests, in-circle tests, or segment intersection. For example, the left-turn predicate is evaluated by computing the sign of the determinant of a 2×2 matrix built from differences of the points' coordinates. Since the evaluation of such a determinant is needed in several other predicates as well, it makes sense to factor out this step into a separate function, which is parameterized by a number type to maintain flexibility even at this level of the kernel. This function can be shared by all predicates and constructions that need to evaluate a 2×2 determinant.

Code reuse is desirable not only because it reduces maintenance overhead but also from a robustness point of view, as it isolates potential problems in a small number of places. Furthermore, these basic numerical operations are equally as accessible to anyone providing additional or customized operations on top of our kernel in the future.

8   Adaptable Algorithms

In the previous sections, we have illustrated the techniques used to realize a kernel concept that includes functors as well as types in a way that makes both

---

[1] beyond a single addition or comparison

easily adaptable. Here we show how such a kernel can be put to good use in the implementation and adaptation of an algorithm.

Kernel as Traits Class In CGAL, the geometric requirements of an algorithm are collected in a geometric traits class which is a parameter of the algorithm's implementation. With the addition of functors to the kernel concept, it is now possible simply to supply a kernel as the argument for the geometric traits class of an algorithm. Consider as a simple example Andrew's variant of Graham's scan [,] for computing the convex hull of a set of points in two dimensions. Assuming the points are already sorted lexicographically, this algorithm requires only a point type and a left-turn predicate from its traits class. Hence, the simple example kernel from Section 4 would suffice.

In general, the requirements of many geometric traits classes are only a subset of the requirements of a kernel. Other geometric traits classes might have requirements that are not part of the kernel concept. They can be implemented as extensions on top, having easy access to the part of their functionality that is provided by the kernel.

Projection Traits As mentioned in Section 5, one benefit of using functors in the traits class and kernel class is the possible association of a state with the functor. This flexibility can be used, for example, to apply a two-dimensional algorithm to a set of coplanar points in three dimensions. Consider the problem of triangulating a set of points on a polyhedral surface. Each face of the surface can be triangulated separately using a two-dimensional triangulation algorithm and a kernel can be written whose two-dimensional part realizes the projection of the points onto the plane of the face in all functors while actually using the original three-dimensional data. The predicates must therefore know about the plane in which they are operating and this is maintained by the functors in a state variable.

Adapting a Predicate Assume, we want to compute the convex hull of a planar point set with a kernel that represents points by their Cartesian coordinates of type `double`[2]. The left-turn predicate amounts to evaluating the sign of a 2×2-determinant; if this is done in the straightforward way by calculations with `doubles`, the result is not guaranteed to be correct due to roundoff errors caused by the limited precision.

By simply exchanging the left-turn predicate, a kernel can be adapted to use a so-called static filter (see also next section) in that predicate. Assume for example, we know that the coordinates of the input points are `double` values from $(-1, 1)$. It can be shown (cf. [ ]) that in this case the correct sign can be determined from the `double` calculation, if the absolute value of the result exceeds $3 \cdot (2^{-50} + 2^{-102}) \approx 2.66 \cdot 10^{-15}$.

9   Kernel Models

The techniques described in the previous sections have been used to realize several models for the geometry kernel concept described in Section 3. In fact, we use class templates to create a whole *family* of models at once. The template

---

[2] A double precision floating point number type as defined in IEEE 754 [ ].

parameter is usually the *number type* used for coordinates and arithmetic. We categorize our kernel families according to *coordinate representation, object reference and construction,* and *level of runtime optimization.* Furthermore, we have actually two kernel concepts in CGAL: a lower-dimensional kernel concept for the fixed dimensions 2 and 3, and a higher-dimensional kernel concept for arbitrary dimension d. For more details beyond what can be presented here, the reader is referred to the CGAL reference manuals [ ].

Coordinate Representation We distinguish between two coordinate representations; Cartesian and homogeneous. The corresponding kernel classes are called `Cartesian<FT>` and `Homogeneous<RT>` with the parameters FT and RT indicating the requirements for a *field type* and *ring type,* respectively. Homogeneous representation allows many operations to factor out divisions into a common denominator, thus avoiding divisions in the computation, which can sometimes improve efficiency and robustness greatly. The Cartesian representation, however, avoids the extra time and space overhead required to maintain the homogenizing coordinate and thus can also be more efficient for certain applications.

Memory Allocation and Construction The standard technique of *smart pointers* can be used to speed up copy constructions and assignments of objects with a reference-counted handle-representation scheme. Runtime experiments show that this scheme pays off for objects whose size is larger than a certain threshold (around 4 words depending on the machine architecture). To allow for an optimal choice CGAL offers for each representation a simple and a smart-pointer based version. In the Cartesian case, these models are called `Simple_cartesian<FT>` and `Cartesian<FT>`.

Filtered Models The established approach for robust geometric algorithms following the exact computation paradigm [ ] requires the exact evaluation of geometric predicates, i.e., decisions derived from geometric computations have to be correct. While this can be achieved straightforwardly by relying on an exact number type [,], this is not the most efficient approach, and the idea of so-called *filters* has been developed to speed up the exact evaluation of predicates [, ,]. See also the example in Section 8. The idea of filtering is to do the calculations on a fast floating point type and maintain an error bound for this approximation. An exact number type is only used where the approximation is not known to give the correct result for the predicate and the hope is that this happens seldom.

CGAL provides an adaptor `Filter_predicate<>`, which makes it easy to use the filter technique for a given predicate, and also a full kernel `Filtered_kernel<>` with all predicates filtered using the scheme presented above.

Higher-Dimensional Kernel The higher-dimensional kernel defines a concept with the same type and functor technology, but is well separated from the lower-dimensional kernel concepts. Higher-dimensional affine geometry is strongly connected to its mathematical foundations in linear algebra and analytical geometry. Since the dimension is now a parameter of the interface and since the solution of linear systems can be done in different ways, a linear algebra concept LA is part of the interface of the higher dimensional kernel models Cartesian_d<FT,LA> and Homogeneous_d<RT,LA>.

10   Conclusions

Many of the ideas presented here have already been realized in CGAL; parts of them still need to be implemented. Although standard compliance is still a big issue for C++ compilers, more and more compilers are able to accept template code such as ours.

We would like to remind the reader that in this paper we have lifted the curtain to how to implement a library, which is considerably more involved than using a library. A user of our design can be gradually introduced to the default use of one kernel, then exchanging one kernel with another kernel in an algorithm, exchanging individual pieces in a kernel, and finally – for experts – writing a new kernel. Only creators of a new library need to know all inner workings of a design, but we believe also interested users will benefit from studying the design.

Finally, note that many topics could be touched very briefly only within the scope of this article. The interested reader will find many more details and examples, in particular regarding the implementation, in the full paper.

Acknowledgments

This work has been supported by ESPRIT LTR projects No. 21957 (CGAL) and No. 28155 (GALIA). The second author also acknowledges support from the Swiss Federal Office for Education and Science (CGAL and GALIA).

Many more people have been involved in the CGAL project, and contributed in one or the other way to the discussion that finally lead to the design presented here. We thank especially Hervé Brönnimann, Bernd Gärtner, Stefan Schirra, Wieger Wesselink, and Mariette Yvinec for their valuable input. Thanks also to Joachim Giesen for comments on the final version.

References

1. ANDREW, A. M. Another efficient algorithm for convex hulls in two dimensions. *Inform. Process. Lett. 9*, 5 (1979), 216–219. 87

2. AUSTERN, M. H. *Generic Programming and the STL.* Addison-Wesley, 1998. 80

3. BAKER, J. E., TAMASSIA, R., AND VISMARA, L. GeomLib: Algorithm engineering for a geometric computing library, 1997. (Preliminary report). 80

4. BARTON, J. J., AND NACKMAN, L. R. *Scientific and Engineering C++.* Addison-Wesley, Reading, MA, 1997. 81

5. BRÖNNIMANN, H., BURNIKEL, C., AND PION, S. Interval arithmetic yields efficient dynamic filters for computational geometry. In *Proc. 14th Annu. ACM Sympos. Comput. Geom.* (1998), pp. 165–174. 88

6. BRÖNNIMANN, H., KETTNER, L., SCHIRRA, S., AND VELTKAMP, R. Applications of the generic programming paradigm in the design of CGAL. In *Generic Programming—Proceedings of a Dagstuhl Seminar* (2000), M. Jazayeri, R. Loos, and D. Musser, Eds., LNCS 1766, Springer-Verlag. 80

7. BURNIKEL, C., MEHLHORN, K., AND SCHIRRA, S. The LEDA class real number. Technical Report MPI-I-96-1-001, Max-Planck Institut Inform., Saarbrücken, Germany, Jan. 1996. 88

8. International standard ISO/IEC 14882: Programming languages – C++. American National Standards Institute, 11 West 42nd Street, New York 10036, 1998. 80, 85

9. CGAL, the Computational Geometry Algorithms Library. http://www.cgal.org/. 79, 80, 88

10. COPLIEN, J. O. Curiously recurring template patterns. C++ Report (Feb. 1995), 24–27. 81

11. DE BERG, M., VAN KREVELD, M., OVERMARS, M., AND SCHWARZKOPF, O. Computational Geometry: Algorithms and Applications. Springer-Verlag, Berlin, 1997. 87

12. FABRI, A., GIEZEMAN, G.-J., KETTNER, L., SCHIRRA, S., AND SCHÖNHERR, S. The CGAL kernel: A basis for geometric computation. In Proc. 1st ACM Workshop on Appl. Comput. Geom. (1996), M. C. Lin and D. Manocha, Eds., vol. 1148 of Lecture Notes Comput. Sci., Springer-Verlag, pp. 191–202. 80

13. FABRI, A., GIEZEMAN, G.-J., KETTNER, L., SCHIRRA, S., AND SCHÖNHERR, S. On the design of CGAL, the computational geometry algorithms library. Software – Practice and Experience 30 (2000), 1167–1202. 80

14. FORTUNE, S., AND VAN WYK, C. J. Static analysis yields efficient exact integer arithmetic for computational geometry. ACM Trans. Graph. 15, 3 (July 1996), 223–248. 88

15. GIEZEMAN, G.-J. PlaGeo, a library for planar geometry, and SpaGeo, a library for spatial geometry. Utrecht University, 1994. 80

16. IEEE Standard for binary floating point arithmetic, ANSI/IEEE Std 754 – 1985. New York, NY, 1985. Reprinted in SIGPLAN Notices, 22(2):9–25, 1987. 87

17. JOSUTTIS, N. M. The C++ Standard Library, A Tutorial and Reference. Addison-Wesley, 1999. 80, 85

18. KARAMCHETI, V., LI, C., PECHTCHANSKI, I., AND YAP, C. The CORE Library Project, 1.2 ed., 1999. http://www.cs.nyu.edu/exact/core/. 88

19. KETTNER, L. Using generic programming for designing a data structure for polyhedral surfaces. Comput. Geom. Theory Appl. 13 (1999), 65–90. 81

20. MEHLHORN, K., AND NÄHER, S. LEDA: A Platform for Combinatorial and Geometric Computing. Cambridge University Press, Cambridge, UK, 2000. 80

21. MUSSER, D. R., AND STEPANOV, A. A. Generic programming. In 1st Intl. Joint Conf. of ISSAC-88 and AAEC-6 (1989), Springer LNCS 358, pp. 13–25. 80

22. MUSSER, D. R., AND STEPANOV, A. A. Algorithm-oriented generic libraries. Software – Practice and Experience 24, 7 (July 1994), 623–642. 80

23. MYERS, N. C. Traits: A new and useful template technique. C++ Report (June 1995). http://www.cantrip.org/traits.html. 81

24. SCHIRRA, S. A case study on the cost of geometric computing. In Proc. Workshop on Algorithm Engineering and Experimentation (1999), vol. 1619 of Lecture Notes Comput. Sci., Springer-Verlag, pp. 156–176. 85

25. SHEWCHUK, J. R. Adaptive precision floating-point arithmetic and fast robust geometric predicates. Discrete Comput. Geom. 18, 3 (1997), 305–363. 87, 88

26. STROUSTRUP, B. The C++ Programming Language, 3rd Edition. Addison-Wesley, 1997. 80

27. VELDHUIZEN, T. Techniques for scientific C++. Technical Report 542, Department of Computer Science, Indiana University, 2000. http://www.extreme.indiana.edu/~tveldhui/papers/techniques/. 85

28. YAP, C. K., AND DUBÉ, T. The exact computation paradigm. In Computing in Euclidean Geometry, D.-Z. Du and F. K. Hwang, Eds., 2nd ed., vol. 4 of Lecture Notes Series on Computing. World Scientific, Singapore, 1995, pp. 452–492. 88

Efficient Resource Allocation with Noisy Functions

Arne Andersson[1], Per Carlsson[2], and Fredrik Ygge[3]

[1] Computing Science Department, Information Technology, Uppsala University
Box 311, SE - 751 05 Uppsala, Sweden
arnea@csd.uu.se
http://www.csd.uu.se/~arnea
[2] Computing Science Department, Information Technology, Uppsala University
Office: Computer Science Department, Lund University
Box 118, SE - 221 00 Lund, Sweden
Per.Carlsson@cs.lth.se
http://www.csd.uu.se/~perc
[3] Enersearch AB and Computing Science Department, Information Technology
Uppsala University, Chalmers Science Park
SE - 412 88 Gothenburg, Sweden, ygge@enersearch.se
http://www.enersearch.se/ygge

Abstract. We consider resource allocation with separable objective functions defined over subranges of the integers. While it is well known that (the maximisation version of) this problem can be solved efficiently if the objective functions are concave, the general problem of resource allocation with functions that are not necessarily concave is difficult.

In this paper we show that for a large class of problem instances with noisy objective functions the optimal solutions can be computed efficiently. We support our claims by experimental evidence. Our experiments show that our algorithm in hard and practically relevant cases runs up to 40 - 60 times faster than the standard method.

1 Introduction

1.1 Resource Allocation

We consider resource allocation with separable objective functions defined over sub-ranges of the integers, as given by

$$
\begin{aligned}
\max_{r_1, r_2, \ldots, r_n} & \quad \sum_{i=1}^{n} f_i(r_i) \\
s.t. & \quad \sum_{i=1}^{n} r_i = R,
\end{aligned}
\tag{1}
$$

where each function $f_i(r_i)$ is defined over a range of integers, I_i.

Resource allocation with separable objective functions is a fundamental topic in optimisation. When presenting algorithms for this problem, it is typically assumed that the objective functions are concave; very little is known about the case of non-concave functions. In the general case, where there is no assumption

G. Brodal et al. (Eds.): WAE 2001, LNCS 2141, pp. 91–105, 2001.
© Springer-Verlag Berlin Heidelberg 2001

on the shape of the functions, the standard method is to use brute force and, for any pair of functions that are aggregated into one, test all possible solutions.

In this paper, we present a new algorithm tailored for the case when there is no prior knowledge about the involved objective functions. The idea is to take advantage of favourable properties of the functions whenever possible. In effect, we obtain a new algorithm for efficient handling of a very general class of non-concave and noisy objective functions.

A typical function with such properties is illustrated in the lower left part of Figure 3. The function is noisy, and even with the noise removed, it is non-concave. The function has some regularity though: seen from a distance it looks "smooth". The basic idea of our algorithm is to divide the objective functions into a small number of intervals and to filter out the noise, such that each interval can be treated as either convex or concave. We then use previous techniques [ ] for efficient pair-wise aggregation of objective functions, extended with a neighbourhood search, as described in this paper. Altogether, we manage to obtain a pruning of the search space; the number of allocations that needs to be tested is significantly reduced. The robustness of our algorithm makes it very useful in practice. Furthermore, the overhead of the algorithm is small enough to allow for fast implementation. Indeed, it competes very well with the brute force algorithm, although the brute force algorithm has the advantage of being simple and straightforward. This is demonstrated by experimental evidence. It is even the case that we achieve surprisingly good results also for seemingly impossible cases. One such case consists of a set of completely random objective functions. At a first glance, it might seem that there is no hope to improve over the brute force method for such a problem, but our new algorithm offers a significant speedup by taking advantage of smoothness whenever possible.

The interest in handling of noisy functions stems from problems that arise in practice *as soon as some of the aggregated functions are non-concave*. We do *not* consider e.g. filtering of random sampling noise on top of the true functions, but treatment of functions that truly have some irregularities. When a number of functions has been aggregated the resulting function is very close to a "smooth" function, but it has irregularities on top. The irregularities originates in the non-concave input functions.

1.2   The Field of the Contribution

The brute force algorithm, as well as our algorithm, is based on pair-wise aggregation of objective functions. In a typical implementation of the brute force method, the global objective function is computed by incorporating the functions one by one [, pp. 47-50], but it is also possible to aggregate the functions in a balanced binary tree fashion [ ]. In our experiments, we have chosen the second alternative mostly because this method is more suited for implementation in a distributed environment, which is prevailing in the application of power load management which we have in mind. We have no reason to believe that the experiments would show any major difference had the other method been used.

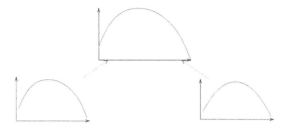

Fig. 1. The basic step when aggregating objective functions is the aggregation of two functions into one

In either case, the basic step is the aggregation of two objective functions into one, see Figure 1. Given two functions f_1 and f_2 defined over subintervals of length I_1 and I_2, compute the aggregated function f_t, defined over a subinterval of length (at most)[1] $I_1 + I_2$:

$$f_t(r) = \max_{r_1+r_2=r} f_1(r_1) + f_2(r_2), \tag{2}$$

and for each r in the subinterval compute the corresponding r_1 and r_2.

The complexity of aggregating two objective functions depends on the properties of the functions. Two cases are typically distinguished, e.g. [,]:

- The general case, when no assumptions can be made about the functions. Then an aggregation requires $\Theta(I_1 I_2)$ time.
- The special case of two concave functions. Then the aggregation can be made much faster, in $\Theta(I_1 + I_2)$ time.

As the domain of the objective functions may be very large compared to the number of functions, the difference between these two complexities will be significant in many applications. Despite the fundamental nature of the problem and its practical importance, we have found nothing applicable done by others in recent literature, cf. [ ], that tackles the problem with lower complexity than what is presented in the textbook by Ibaraki and Katoh [ ], i.e. $\Theta(I_1 I_2)$.

In previous work [ ], we have introduced an algorithm for pair-wise aggregation of objective functions. The main point of that work is that the complexity of the aggregation is adaptive to the shape of the two functions to be aggregated; the simpler curves the lower complexity.

The main contribution of this paper is that it generalises the previous work to manage noisy cases. First, we are able to handle functions that are close to

---

[1] If the total available resource to be allocated is smaller than $I_1 + I_2$, then the aggregated function need not be computed over the entire interval $I_1 + I_2$.

concave but noisy (see Figure 2) in an efficient way. Furthermore, we handle noisy functions that basically are non-concave but in some sence smooth (as the ones in Figure 3).

It is worth pointing out that noisy looking objective functions do not only occur when the input functions are noisy themselves; even with nicely shaped input functions the functions occurring after a number of aggregations are functions that are close to "smooth" functions but are irregular. The irregularities originate from non-concave segments of input functions, see the experimental results in Section 4 and Figure 7.

Another important contribution is our implementation work and the experiments showing the robustness of our algorithm.

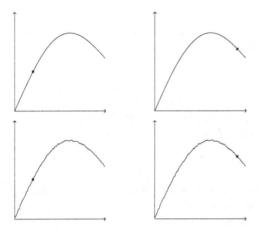

Fig. 2. Top row: The allocation of $r = r_1 + r_2$ indicated by the two dots is obviously not optimal and it is well known how to avoid testing it for optimality. Bottom row: It is easy to see that the allocation is not optimal also when the functions are slightly noisy and it is rather easy to avoid testing it for optimality

The full description of the algorithm is rather extensive and cannot be given in this paper, but is available in a technical report [ ]. In addition, it is possible to download the Java classes needed to run the test program, download site: http://www.csd.uu.se/~perc/papers/noisy/.

Outline of the paper: In Section 2 we state our main result. In Section 3 we give a high level overview of the algorithm. This is followed, in Section 4, by experimental results, and finally we give some conclusions, Section 5.

2   Main Result

In this work we show that it is possible to aggregate two objective functions efficiently and optimally if the objective functions could be divided into a small

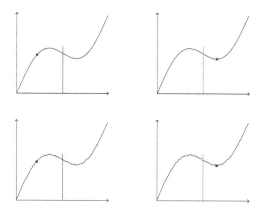

Fig. 3. Top row: In previous work [ ] we have shown how to efficiently compute the optimal allocation when the objective functions are not restricted to concave functions. The basic idea is to divide the functions into segments. Bottom row: In this paper we turn to the practically relevant case when the objective functions can be accurately approximated with fairly well shaped functions. By filtering out the noise the functions can be partitioned into segments as the functions in the top row. This is combined with a neighbourhood search to obtain the optimal solution. Also in this figure it is clear that the marked allocations can not be optimal, though constructing an efficient algorithm for the bottom case is non-trivial

number of segments that essentially are concave and convex but have some rather low amplitude noise (see Figure 2 and 3). We present an algorithm, Algorithm 1, that is based on the algorithm presented in our previous work, [ ], but is generalised to manage low amplitude noise.

It is well known, [ ], [ ], that it is possible to aggregate two functions by testing all combinations. (Here referred to as the brute force method.)

Statement 1. *Let f_1 and f_2 be two objective functions (not necessarily concave) defined over intervals of length I_1 and I_2. The aggregated function and the corresponding allocations can be computed in $\mathcal{O}(I_1 \cdot I_2)$ time by testing of all combinations [ ].*

With n objective functions to aggregate this can be done in $\mathcal{O}(nR^2)$ time, where R is the resource to be allocated.

We have shown that when it is possible to divide the functions into a (preferably small) number of segments, cf. Figure 3, top row, that are either concave or convex it is possible to do better.

The algorithm is a two step algorithm. First the functions are divided into segments (in linear time), then all segments of the first function are combined with all segments of the second one in a search for candidate allocations. We have shown that, for each segment combination, this search can be done in time linear in the size of the segments.

Statement 2. *Let f_1 and f_2 be two objective functions defined over intervals of length I_1 and I_2. Furthermore, assume that the two functions can be divided into s_1 and s_2 segments respectively, such that each segment is either convex or concave. Then the aggregated objective function and the corresponding optimal allocations can be computed in $\mathcal{O}(I_1 s_2 + I_2 s_1)$ time [ ].*

In some situations the growth of the complexity as a function of the number of segments turns out to be a problem since the number of segments often grows large due to low amplitude noise. This noise arises when (some of) the functions are non concave, cf. Section 4. A consequence is that the overall complexity grows and reaches the complexity of testing all combinations.

A way of solving this is to filter out the noise by accepting a small hull distance (defined in Section 3) when dividing the functions into (concave and convex) segments. Then the neighbourhood of each candidate allocation is searched until the (verified) locally optimal solution has been found. The algorithm combines all segments of the first function with all segments of the second one. Since a globally optimal allocation has to be locally optimal for some segment combination, all globally optimal allocations are found this way.

Statement 3. *Let f_1 and f_2 be two objective function (not necessarily concave), with relatively small amplitude noise. Furthermore, assume that after a small adjustment of each function value, the functions could be divided into a small number of segments, such that each segment is either convex or concave.*

Then the aggregated objective function and the corresponding optimal allocations can be computed in time considerably less than what is needed for a complete search of all possible combinations.

The complexity is dependent on the number of segments the functions are divided into and the amplitude of the noise.

However, with n objective functions to aggregate we still have not managed to achieve an overall complexity lower than $\mathcal{O}(nR^2)$. Still, the algorithm offers a pruning of the search space (compared to a complete search of all combinations) and running times that in practical cases are considerably less than the running times of a the brute force algorithm (see Section 4).

3   Technical Overview

Due to space constraints we can only give a very high level descriptions here, and we refer to our technical report [ ] for details.

The problem of noisy functions described above is solved with an algorithm whose basic idea can be described in the following high level way :

(i) Divide the functions to be aggregated into segments that with a given (preferably small) hull distance could be viewed as convex and concave, and accurately describes the functions, see Figure 4, and

(ii) follow our previous algorithm (designed without the idea of a hull distance) [ ] on the hull functions (that are either concave or convex) over the segments but run a few extra steps in a neighbourhood search to guarantee that the solution is optimal for the original functions. (Hull functions and hull distances are defined below.)

Hull functions and hull are defined as follows:

Definition 1. $\hat{f}$ *is the smallest concave function such that* $\hat{f}(r) \geq f(r)$ *for all* r *within a specific segment.*
In the same way $\check{f}$ *is the largest function that is convex such that* $\check{f}(r) \leq f(r)$ *for all* r *within the segment.*

The functions $\hat{f}$ and $\check{f}$ are illustrated in Figure 4. We use $\hat{f}$ and $\check{f}$ as functions guiding the search. In this way the number of segments that the function is divided into is reduced.

Definition 2. *The* hull function *of a segment is either* $\hat{f}$ *or* $\check{f}$.

Definition 3. *We define the* hull distance, ϵ, *as the maximum distance between* f *and* $\hat{f}$ *or* $\check{f}$ *(depending on which one is used as hull function of the segment), see Figure 4.*

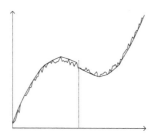

Fig. 4. A noisy function is divided into segments with a hull function that is either concave or convex

3.1   The Algorithm

Once again, the algorithm is a two step algorithm. First the functions are divided into segments, Section 3.2. Then our previous algorithm and a neighbourhood search strategy is applied on all possible segment combinations. For a pseudo code representation of the algorithm this has to be expressed more formally.

Recall the definitions of f_1, f_2, f_t, I_1 and I_2 from Equation (2). Let the aggregated function, f_t, be represented by a vector $\mathbf{f}_t$, and let the corresponding (optimal) allocations of r_1 and r_2 for each r be represented by the vectors $\mathbf{r}_1$ and $\mathbf{r}_2$ respectively.[2]

With this notation we can express our algorithm for pair-wise aggregation of objective functions as follows:

Algorithm 1. `Algorithm for aggregation of two functions`
`For every element, $i$, of the $\mathbf{f}_t$ vector`
   $\mathbf{f}_t(i) \leftarrow -\infty$
`Divide $f_1$ and $f_2$ into segments with concave and convex hull functions`
`For every combination of segments with one segment from $f_1$ and one from $f_2$`
   `For every local candidate allocation, $(r_1, r_2)$`
   `of the hull functions`
      {
         `Search the neighbourhood`[3]
         `If $\mathbf{f}_t(r'_1 + r'_2) < f_1(r'_1) + f_2(r'_2)$`
         `for any $(r'_1, r'_2)$ of`
         `the neighbourhood of $(r_1, r_2)$`
         {
            $\mathbf{f}_t(r'_1 + r'_2) \leftarrow f_1(r'_1) + f_2(r'_2)$
            $\mathbf{r}_1(r'_1 + r'_2) \leftarrow r'_1$
            $\mathbf{r}_2(r'_1 + r'_2) \leftarrow r'_2$
         }
      }

In the following two sections we give a brief overview of the two steps of the algorithm, the construction of segments (that is not the main focus of our work) and the actual search for optimal allocations.

3.2   Constructing the Segments

The segments could be constructed in several ways. Here we give a short description of the strategy used in the test implementation.

The segments are constructed using the incremental algorithm for building the convex hull of a point set, see e.g. [ ], applied on (a part of) the objective function. This gives a lower path equal to $\check{f}$ and an upper path equal to $\hat{f}$.

This is done in a recursive fashion. First construct a hull function on the entire function. If the hull distance is not bigger than tolerated for either $\check{f}$ or $\hat{f}$ we are done. Otherwise divide the segment in equal halves and repeat the process.

3.3   Finding all Candidate Allocations

All segments of the first function are combined with all segments of the second function in a search for candidate allocations. The main principle of the

---

[2] Again, if the total resource which can be allocated is smaller than $I_1 + I_2$, then smaller vectors can be used.

[3] Described in the lemmas of our technical report.

algorithm is to apply the old algorithm [ ] to the hull functions and use the candidates as starting points for a neighbourhood search for the optimal solution on f_1 and f_2. In this way all possible optimal allocations of the true functions are found.

There are three possible segment combinations:

- two segments with concave hull functions,
- two segments with convex hull functions, and
- one segment with a concave hull function and one with a convex hull function.

Defining the neighbourhoods of the first two combinations is rather straight forward, and for the third one it is hard [ ]. For illustration we sketch the principles of the neighbourhood search for the first combination, two segments with concave hull functions. The idea of the neighbourhood definition could be viewed in Figure 5. Loosely we could say that when the combination of the hull functions is less valuable than the best allocation found on the true functions the search could be pruned.

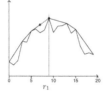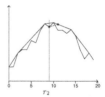

Fig. 5. The allocation $r_1 = r_2 = 9$, which is optimal with respect to the hull functions, is the starting point of a neighbourhood search. The search is ended when the slope of the dashed line with diamonds of the left pane is greater than the corresponding slope in the right pane

Since all segment combinations are searched all optimal allocations are found [ ].

4   Experimental Results

We have run a series of tests and the results are illuminating. The tests were run using the algorithm in a tree structured [ ] system for resource allocation. The system is implemented in Java and the tests that are referred below were run on a Sun Ultra 10 (tests have been run on other Java systems too with small variations in the results).

4.1   Tree-Structured Aggregation

As pointed out above, the basic step of solving the resource allocation problem is the aggregation of two objective functions into a new objective function. In our experiments, we have used this basic step within a tree structure. As leaves of the tree, we have the original objective functions, the internal nodes are aggregated functions, the global objective function being the root, cf. Figure 6. One great advantage of this is that it is highly suitable for distributed resource allocation. With this structure it is possible to distribute the computational work in the network and to avoid communication bottlenecks, since all the computation does not have to be done at one central point [ ].

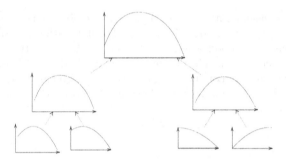

Fig. 6. We have presented an algorithm for resource allocation that is designed with distributed resource allocation in mind, [ ]. In this paper we are focusing the pair-wise aggregation that is an essential subroutine of this and other algorithms

4.2   Experiments

In our tests[4] our algorithm for pair-wise aggregation is used within an algorithm that

(i)   is tree structured (Figure 6),
(ii) in the bottom, where the functions are defined over small intervals, performs a complete search of all combinations, like the brute force algorithm,
(iii) uses our previous algorithm until the number of segments is considered too large, and
(iv) when the number of segments is considered too large turns to using the algorithm introduced in this paper.

The main point of this section is to show that our novel algorithm not only prunes part of the search space compared to the more simple algorithm of testing all combinations, but also is faster for practically relevant instances (despite its larger overhead).

---

[4] The Java classes needed for running the tests, and a couple of other small test programs, can be downloaded from http://www.csd.uu.se/~perc/papers/noisy/.

Timing Considerations In our experiments we have measured the total running time as well as the time and number of evaluations used at the top level. The aggregation at the lower levels of the aggregation hierarchy is of minor interest (the functions are typically defined over small intervals and a complete search is of low cost). Therefor we focus on the top level measure in our evaluation (see the tables below).

Experiment 1: Two Functions Based on a Set of Five Hundred Mixed Functions The first example is chosen to reflect our application area of power load management. It is realistic to assume that the objective functions of the vast majority of the loads in an electricity grid are concave. However, some functions will be non-concave, e.g. staircase shaped. A few levels up in the aggregation hierarchy the aggregated objective functions could be accurately approximated with a concave function, but a low amplitude noise prevents our previous aggregation algorithm from being highly efficient, see Figure 7. Also, because of this noise, standard algorithms for concave functions cannot be applied.

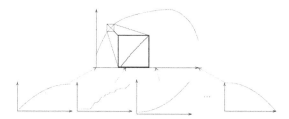

Fig. 7. The resulting function when one hundred mixed functions are aggregated is almost concave. (A part of the aggregated function is enlarged so that it is possible to see the noise)

The functions that we aggregate in this test are five hundred functions that are randomly generated and ordered. One third of them are concave, one third are convex, and one third are staircase shaped. The length of the intervals that the functions are defined over differs up to a factor five. All variations are randomly distributed over the set of functions. The number of segments accepted before trying to reduce it is chosen so that the algorithm for pair-wise aggregation based on hull functions is used at higher levels of the aggregation hierarchy. A series of tests were run with different hull distances (which effect the number of segments and the neighbourhood searched), as described in Table 1.

With this input and configuration the aggregation at the top level was between three and 44 times faster with our new algorithm compared to the simple algorithm of Statement 1. The variance is due to the choice of hull distance when constructing the segments, see Section 3.2. Counting actual comparisons the difference is even bigger.

Table 1. Aggregating 500 mixed functions. Our algorithm for pair-wise aggregation compared to a complete search of all possibilities. The ϵ is the hull distance (defined in Section 3.2). In this case the running time of the top level aggregation becomes independent of the hull distance when $\epsilon \geq 4$ since the algorithm is able to construct one concave hull function per function, cf. Section 3.2

Instance	Total Time	#Evaluations (top level)	Time (s) (top level)
Complete Search	72.1	61,868,250	33.2
Our Alg., $\epsilon = 0.01$	33.3	8,692,049	11.5
Our Alg., $\epsilon = 0.1$	12.3	890,327	2.4
Our Alg., $\epsilon = 0.25$	10.4	602,657	1.4
Our Alg., $\epsilon = 0.5$	9.8	557,067	1.2
Our Alg., $\epsilon = 1$	9.6	666,945	1.4
Our Alg., $\epsilon = 2$	9.9	621,584	1.3
Our Alg., $\epsilon = 4$	9.3	482,156	0.7

Experiment 2: A Noisy Concave Function and a Smooth Function
Often the function that is noisy could be accurately approximated with a single concave function, see Figure 2 and 7. Although this is the normal case, there are situations, e.g. in power load management, where this does not hold.

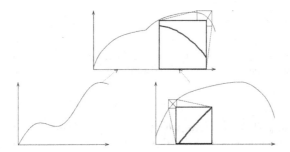

Fig. 8. If e.g. a power plant with a staircase shaped objective function is aggregated with a noisy concave function the result is a non concave noisy function. (Parts of the noisy functions are enlarged so that it is possible to see the noise)

One could think of a power plant that has two generators, and thereby an objective function that is staircase shaped (as the one in Figure 8, bottom left). Due to its large impact on the system it is introduced on a rather high level in the aggregation hierarchy (so that it is in balance with the other nodes on the same level). When a staircase shaped function is aggregated with a noisy concave function the result often is non concave and noisy, cf. Figure 8.

Table 2. Aggregating a staircase shaped function, e.g. the objective function of a power plant, and a noisy concave function. Our algorithm for pair-wise aggregation compared with a complete search of all possibilities. The ϵ is the hull distance (defined in Section 3.2)

Instance	#Evaluations	Time (s)
Complete Search	8,865,000	4.83
Our Alg., $\epsilon = 0.01$	874,046	1.66
Our Alg., $\epsilon = 0.1$	307,232	0.56
Our Alg., $\epsilon = 0.25$	357,254	0.52
Our Alg., $\epsilon = 0.5$	479,622	0.92
Our Alg., $\epsilon = 1$	558,127	0.67
Our Alg., $\epsilon = 2$	794,694	0.95

The functions used in this test was a staircase shaped function (e.g. the objective function of a power plant) and a noisy function that was close to concave (a function that was the result of an aggregation of a set of mixed functions, as the set in experiment one). The functions were defined over around 3,000 sample points each. As described in Table 2 our algorithm was between three and ten times faster than the brute force algorithm.

Experiment 3: Two Functions Based on a Set of Five Hundred Random Functions As an adversary test based on a larger set of input functions we have tried random noise functions, see Figure 9. However, even here the algorithm behaves surprisingly well.

With a set of five hundred functions constructed with random values in $[0 \ldots 1]$ and a reasonable choice of hull distance the algorithm runs the top level aggregation up to 65 times faster than the complete search algorithm, see Table 3.

Table 3. Aggregating 500 random functions. Our algorithm for pair-wise aggregation compared with a complete search of all possibilities. The ϵ is the hull distance (defined in Section 3.2). The running time of the top level aggregation becomes independent of the hull distance when $\epsilon \geq 0.1$ since the algorithm is able to construct one concave hull function per function, cf. Section 3.2

Instance	Total Time	#Evaluations (top level)	Time (s) (top level)
Complete Search	69.9	62,115,388	38.6
Our Alg., $\epsilon = 0.01$	28.4	893,310	2.45
Our Alg., $\epsilon = 0.1$	6.68	353,444	0.59

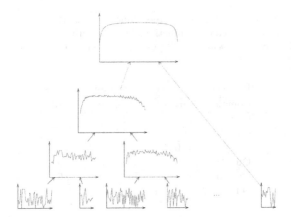

Fig. 9. The resulting function when five hundred random functions are aggregated is almost concave. Already at the second aggregation level the almost concave shape of the aggregated function is obvious

4.3 Summary of the Experiments

Our new algorithm not only theoretically prunes part of the search space, but also runs faster than the more simple method of testing all combinations (despite its larger overhead). It should also be noted that whereas the implementation of the brute force algorithm can be expected to be close to optimised (because of its simplicity) much could probably be done to improve our first test implementation of the more complicated algorithm, particularly regarding the segmentation (which is not in focus in this paper).

5   Conclusions

In this paper, we have presented a robust algorithm for resource allocation with separable objective functions. It is particularly interesting to note that we can handle seemingly impossible instances, such as random functions, more efficient than the obvious brute force algorithm. Simply said, whenever there is some regularity in the input, either in the original objective functions or in the intermediate aggregated functions, we take advantage of this.

Although our algorithm is more involved than the brute force algorithm, the implementation overhead is affordable, as shown by our experiments. Therefore, our new algorithm is a competitive candidate for practical applications.

References

[1] A. Andersson, P. Carlsson, and F. Ygge. Resource allocation with noisy functions. Technical Report 2000-017, Department of Information Technology, Uppsala Uni-

versity, July 2000. (Available from www.it.uu.se/research/reports/). 94, 96, 99

[2] A. Andersson and F. Ygge. Managing large scale computational markets. In H. El-Rewini, editor, *Proceedings of the Software Technology track of the 31th Hawaiian International Conference on System Sciences (HICSS31)*, volume VII, pages 4–14. IEEE Computer Society, Los Alamos, January 1998. ISBN 0-8186-8251-5, ISSN 1060-3425, IEEE Catalog Number 98TB100216. (Available from http://www.enersearch.se/ygge). 92, 95, 99, 100

[3] Arne Andersson and Fredrik Ygge. Efficient resource allocation with non-concave objective functions. *Accepted for publication in Computational Optimization and Applications*, 2001. (A preprint version is available as a research report from http://www.enersearch.se). 92, 93, 95, 96, 97, 99

[4] Thomas H. Cormen, Charles E. Leiserson, and Ronald L. Rivest. *Introduction to Algorithms*. MIT Press, Cambridge, Massachusetts London, England, 1989. 98

[5] R. Fletcher. *Practical Methods of Optimization*. John Wiley & Sons, 1987. Second Edition. 93

[6] R. Horst, P. Pardalos, and N. Thoai. *Introduction to Global Optimization*. Kluwer, Norwell, MA, 1995. 93

[7] Tochihide Ibaraki and Naoki Katoh. *Resource Allocation Problems — Algorithmic Approaches*. The MIT Press, Cambridge, Massachusetts, 1988. 92, 93, 95

Improving the Efficiency of Branch and Bound Algorithms for the Simple Plant Location Problem

Boris Goldengorin, Diptesh Ghosh, and Gerard Sierksma

Faculty of Economic Sciences, University of Groningen
P.O. Box 800, 9700AV Groningen, The Netherlands
{B.Goldengorin,D.Ghosh,G.Sierksma}@eco.rug.nl

Abstract. The simple plant location problem is a well-studied problem in combinatorial optimization. It is one of deciding where to locate a set of plants so that a set of clients can be supplied by them at the minimum cost. This problem often appears as a subproblem in other combinatorial problems. Several branch and bound techniques have been developed to solve this problem. In this paper we present some techniques that enhance the performance of branch and bound algorithms. Computational experiments show that the new algorithms thus obtained generate less than 60% of the number of subproblems generated by branch and bound algorithms, and in certain cases require less than 10% of the execution times required by conventional branch and bound algorithms.

1 Introduction

Given sets $I = \{1, 2, \ldots, m\}$ of sites in which plants can be located, $J = \{1, 2, \ldots, n\}$ of clients, a vector $F = (f_i)$ of fixed costs for setting up plants at sites $i \in I$, a matrix $C = [c_{ij}]$ of transportation costs from $i \in I$ to $j \in J$, and an unit demand at each client site, the Simple Plant Location Problem (SPLP) is the problem of finding a set S, $\emptyset \subset S \subseteq I$, at which plants can be located so that the total cost of satisfying all client demands is minimal. The costs involved in meeting the client demands include the fixed costs of setting up plants, and the transportation cost of supplying a given client from an open plant. The SPLP forms the underlying model in several combinatorial problems, like set covering, set partitioning, information retrieval, simplification of logical Boolean expressions, airline crew scheduling, vehicle despatching (Christofides [ ]), assortment (Beresnev *et al.* [ ], Goldengorin [ ], Jones *et al.* [ ], Pentico [ ], Tripathy *et al.* [ ]) and is a subproblem for various location analysis problems (Revelle and Laporte [ ]). The SPLP is NP-hard (Cornuejols *et al.* [ ]), and many exact and heuristic algorithms to solve the problem have been discussed in the literature. Most of the exact algorithms are based on a mathematical programming formulation of the SPLP. Interestingly, none of them explicitly incorporate preprocessing. A detailed discussion of exact algorithms for the SPLP appears in Cornuejols *et al.* [ ].

G. Brodal et al. (Eds.): WAE 2001, LNCS 2141, pp. 106–117, 2001.
© Springer-Verlag Berlin Heidelberg 2001

The pseudo-Boolean representation of the SPLP facilitates the construction of rules to reduce the size of SPLP instances (Beresnev *et al.* [ ], Cornuejols *et al.* [ ], Dearing *et al.* [ ], Goldengorin *et al.* [ ], etc.). These rules have never been used in the explicit description of any algorithm for the SPLP in the available literature. In this paper we propose enhancements to branch and bound algorithms that use these rules, not only for preprocessing, but also as a tool to either solve a subproblem or reduce its size. We also use coefficients of a pseudo-Boolean polynomial representation of the SPLP to compute efficient branching functions. For the sake of simplicity, we use a common depth first branch and bound scheme in our implementations and a simple combinatorial bound, but the concepts developed herein can easily be implemented in any of the algorithms cited in Cornuejols *et al.* [ ]. Our implementations demonstrate the effectiveness of these enhancements for branch and bound algorithms.

The remainder of this paper is organized as follows. In Section 2 we describe a pseudo-Boolean representation of the SPLP. In Section 3 we describe the proposed enhancements in detail, and in Section 4, our computational experience with them. We summarize the paper in Section 5 and propose directions for further research.

2   A Pseudo-Boolean Approach to SPLP

The pseudo-Boolean approach to solving the SPLP (Hammer [ ], Beresnev [ ]) is based on the fact that any instance of the SPLP has an optimal solution in which each client is supplied by exactly one plant. This implies, that in an optimal solution, each client will be served fully by the open site closest to it. Therefore, it is sufficient to determine the sites where plants are to be located, and then use a minimum weight matching algorithm to assign clients to plants. We present below a combinatorial (1) and a pseudo-Boolean (5) statement of the SPLP.

An instance of the SPLP can be described by a m-vector $F = (f_i)$, and a $m \times n$ matrix $C = [c_{ij}]$. We assume that elements of F and C are nonnegative. The total cost $f_{[F|C]}(S)$ associated with a solution S consists of two components, the fixed costs $\sum_{i \in S} f_i$, and the transportation costs $\sum_{j \in J} \min\{c_{i,j} | i \in S\}$, i.e. $f_{[F|C]}(S) = \sum_{i \in S} f_i + \sum_{j \in J} \min\{c_{i,j} | i \in S\}$, and the SPLP is the problem of finding

$$S^\star \in \arg\min\{f_{[F|C]}(S) : \emptyset \subset S \subseteq I\}. \tag{1}$$

A $m \times n$ *ordering matrix* $\Pi = [\pi_{ij}]$ is a matrix each of whose columns $\Pi_j = (\pi_{1j}, \ldots, \pi_{mj})^T$ define a permutation of $1, \ldots, m$. Given a transportation matrix C, the set of all ordering matrices Π such that $c_{\pi_{1j}j} \leq c_{\pi_{2j}j} \leq \cdots \leq c_{\pi_{mj}j}$, for $j = 1, \ldots, n$, is denoted by $perm(C)$. Defining

$$y_i = \begin{cases} 0 & \text{if } i \in S \\ 1 & \text{otherwise,} \end{cases} \quad \text{for each } i = 1, \ldots, m \tag{2}$$

we can indicate any solution S by a vector $\mathbf{y} = (y_1, y_2, \ldots, y_m)$. The fixed cost component of the total cost can be written as $\mathcal{F}_F(\mathbf{y}) = \sum_{i=1}^{m} f_i(1 - y_i)$.

Given a transportation cost matrix C, and an ordering matrix $\Pi \in perm(C)$, we can denote differences between the transportation costs for each $j \in J$ as

$$\Delta c[0, j] = c_{\pi_{1j}j}, \quad \text{and}$$
$$\Delta c[l, j] = c_{\pi_{(l+1)j}j} - c_{\pi_{lj}j}, \quad l = 1, \ldots, m - 1.$$

Then, for each $j \in J$, $\min\{c_{i,j} | i \in S\} = \Delta c[0, j] + \Delta c[1, j] \cdot y_{\pi_{1j}} + \Delta c[2, j] \cdot y_{\pi_{1j}} \cdot y_{\pi_{2j}} + \cdots + \Delta c[m - 1, j] \cdot y_{\pi_{1j}} \cdots y_{\pi_{(m-1)j}} = \Delta c[0, j] + \sum_{k=1}^{m-1} \Delta c[k, j] \cdot \prod_{r=1}^{k} y_{\pi_{rj}}$, so that the transportation cost component of the cost of a solution $\mathbf{y}$ corresponding to an ordering matrix $\Pi \in perm(C)$ is $T_{C,\Pi}(\mathbf{y}) = \sum_{j=1}^{n} \left\{ \Delta c[0, j] + \sum_{k=1}^{m-1} \Delta c[k, j] \cdot \prod_{r=1}^{k} y_{\pi_{rj}} \right\}$.

Thus the total cost of a solution $\mathbf{y}$ to the instance $[F|C]$ corresponding to an ordering matrix $\Pi \in perm(C)$ is given by the pseudo-Boolean polynomial

$$f_{[F|C],\Pi}(\mathbf{y}) = \mathcal{F}_F(\mathbf{y}) + T_{C,\Pi}(\mathbf{y})$$
$$= \sum_{i=1}^{m} f_i(1 - y_i) + \sum_{j=1}^{n} \left\{ \Delta c[0, j] + \sum_{k=1}^{m-1} \Delta c[k, j] \cdot \prod_{r=1}^{k} y_{\pi_{rj}} \right\}. \quad (3)$$

It can be shown (Goldengorin et al. [ ]) that the total cost function $f_{[F|C],\Pi}(\cdot)$ is identical for all $\Pi \in perm(C)$. We call this pseudo-Boolean polynomial the Beresnev function $\mathcal{B}_{[F|C]}(\mathbf{y})$ corresponding to the SPLP instance $[F|C]$ and $\Pi \in perm(C)$. In other words

$$\mathcal{B}_{[F|C]}(\mathbf{y}) = f_{[F|C],\Pi}(\mathbf{y}) \text{ where } \Pi \in perm(C). \quad (4)$$

We can formulate (1) in terms of Beresnev functions as

$$\mathbf{y}^* \in \arg\min\{\mathcal{B}_{[F|C]}(\mathbf{y}) : \mathbf{y} \in \{0, 1\}^m, \mathbf{y} \neq \mathbf{1}\}. \quad (5)$$

Beresnev functions assume a central role in the enhancements described in the next section.

3   Enhancing Branch and Bound Algorithms

In this section we describe enhancements to Branch and Bound (BnB) algorithms to solve SPLP instances. The algorithms thus obtained are collectively called *branch and peg* (BnP) algorithms. We will use depth first search in our algorithms, but the concepts can also be used unaltered in best first search. We use the following notation in the remainder of this section. A *solution* to a SPLP instance with $|I| = m$ and $|J| = n$ is denoted by a vector $\mathbf{y}$ of length m, that is defined on the alphabet $\{0, 1, \#\}$. $y_j = 0$ (respectively, $y_j = 1$) indicates that a plant will be located (respectively, not located) at the site with index j. $y_j = \#$

indicates a decision regarding locating a plant at the site with index j has not yet been reached. A vector $\mathbf{y}$ with $y_j = \#$ for some j is called a *partial solution*, while all other solutions are called *complete solutions*. The process of setting the value of y_j for some index j in any partial solution $\mathbf{y}$ to 0 or 1 is called *pegging*. Indices j for which $y_j = \#$ are called *free* indices while the other indices (for which $y_j = 0$ or 1) are called *pegged* indices.

The Beresnev function corresponding to the SPLP allows us to develop rules using which we can peg certain indices in a solution by examining the coefficients of the terms in it. The rule that we use here was first described in Goldengorin *et al.* [ ] as a preprocessing rule.

Pegging Rule (Goldengorin *et al.* [ ]) *Let $\mathcal{B}_{[F|C]}(\mathbf{y})$ be the Beresnev function corresponding to a SPLP instance $[F|C]$ in which like terms have been aggregated. Let a_k be the coefficient of the linear term corresponding to y_k and let t_k correspond to the sum of all non-linear terms containing y_k. Then*

(a) *If $a_k \geq 0$ then there is an optimal solution $y^\star$ in which $y_k^\star = 0$.*
(b) *If $a_k < 0$, $|a_k| \geq t_k$, and setting $y_k = 1$ does not render the current partial solution infeasible, then there is an optimal solution $y^\star$ in which $y_k^\star = 1$.*

The key factor in the implementation of this pegging rule is an efficient way of computing a_k and t_k values for each index in a partial solution. We implemented this using the ordering matrix Π and two vectors, *top* and *bottom*. Initially, for each $j \in J$, $top_j = 2$ and $bottom_j = m$. Figures 1–3 describe the computation of a_k and t_k for any index k using *top* and *bottom*, and updating of *top* and *bottom* after pegging of any index.

```
function compute a_k(k: index)
begin
    set a_k ← -f_i;
    for each index j ∈ J
    begin
        if top_j points to an element in a row lower than that
        pointed to by bottom_j
            continue;
        if top_j points to an entry not marked k
            continue;
        a_k ← a_k + c_{π_top_j j} - c_{π_top_j-1 j};
    end;
    return a_k;
end;
```

Fig. 1. Computing a_k

In the remainder of this section, we will assume the existence of a function *PegPartialSolution* that takes in a partial solution as its input and returns the solution obtained by applying the pegging rule repeatedly until no variable

```
function compute tₖ(k: index)
begin
    set tₖ ← 0;
    for each index j ∈ J
    begin
        if topⱼ points to an element in a row lower than that
        pointed to by bottomⱼ
            continue;
        t ← m such that πₘ ⱼ = k;
        tₖ ← tₖ + c_{π_{bottomⱼ j}} − c_{π_{t j}};
    end;
    return tₖ;
end;
```

Fig. 2. Computing t_k

```
function update (k: index)
begin
    if k is pegged to 1
    begin
        for each index j ∈ J
        begin
            t ← m such that m > topⱼ, y_{πₘ j} ≠ 1;
            topⱼ ← m;
        end;
    end
    else if k is pegged to 0
    begin
        for each index j ∈ J
        begin
            t ← m such that πₘ ⱼ = k;
            bottomⱼ ← min{t, bottomⱼ};
        end;
    end
end;
```

Fig. 3. Updating *top* and *bottom*

could further be pegged. This rule will be used for preprocessing in both BnB and BnP algorithms, as well as for pegging variables in partial solutions in the BnP search tree. For notational convenience, we will denote the solution obtained after preprocessing, i.e. by running *PegPartialSolution* on $(\#\# \cdots \#)$, the *initial solution*. This solution forms the root of the BnP and BnB search trees.

The choice of the variable to branch on is critical for the success of a branch and bound scheme. The following trivial branching function can be used in the absence of any prior knowledge regarding the suitability of the variable to branch on.

Branching Function 1 *Return the variable with the minimum free index in the subproblem.*

However, we could use information from the coefficients of the Beresnev function to create a more effective branching functions.

Consider a subproblem in which the partial solution, after being pegged by the *PegPartialSolution* function, is $\mathbf{y}$. For each variable y_k in the solution that has not been pegged, let $\mathbf{y}^{k0}$ be obtained by forcibly pegging y_k to 0, and running *PegPartialSolution* on the resultant solution; and let ϕ_{k0} be the number of free indices in $\mathbf{y}^{k0}$. Similarly, let $\mathbf{y}^{k1}$ be obtained by forcibly pegging y_k to 1 in $\mathbf{y}$, and running *PegPartialSolution* on the resultant solution; and let ϕ_{k1} be the number of free indices in $\mathbf{y}^{k1}$. If we want to obtain a quick upper bound for the solution at the current subproblem by solving its subproblems by pegging, then $\phi_k = \min(\phi_{k0}, \phi_{k1})$ is a good measure of the suitability of y_k as a branching variable. (Other combinations of ϕ_{k0} and ϕ_{k1}, such as $\frac{\phi_{k0}+\phi_{k1}}{2}$ could also be used, but our preliminary experimentation shows that these do not cause significant differences in the results obtained.) A branching function based on such a measure can be expected to generate relatively fewer subproblems while solving a SPLP instance. However, the calculations involved would take excessive time. As a compromise therefore, we could use a branching function that generates the ordering of the indices once for the initial solution and uses it for all subproblems. This branching function is described below.

Branching Function 2 *For a free index k in the initial solution $\mathbf{y}$, let ϕ_{k0} (respectively, ϕ_{k1}) be the number of free indices in the solution obtained by setting $y_k = 0$ (respectively, $y_k = 1$) in the initial solution and running PegPartial-Solution on it. Define the fitness of ϕ_i the index i as*

$$\phi_i = \begin{cases} \min(\phi_{i0}, \phi_{i1}) & \text{if } i \text{ is free in the initial solution,} \\ \infty & \text{otherwise.} \end{cases}$$

Generate an ordering of indices $1, \ldots, m$, such that index p precedes index q only if $\phi_p \leq \phi_q$. Return the variable with the minimum free index in this ordering.

A third branching function may be devised in the following manner. Consider a subproblem in which the partial solution, after being pegged by the *PegPartialSolution* function, is $\mathbf{y}$. From the Pegging Rule, we conclude that

$a_k < 0$ and $t_k + a_k > 0$ for each variable y_k in the solution that has not been pegged. y_k would have been pegged to 0 in this solution if the coefficient of linear term involving y_k in the Beresnev function would have been increased by $-a_k$. It would have been pegged to 1, if the same coefficient would have been decreased by $t_k + a_k$. Therefore we could use $\phi_k = \max(-a_k, t_k + a_k)$ as a measure of the improbability of y_k being pegged in any subproblem of the current subproblem. If we want to reduce the size of the branch and bound tree by pegging such variables, then we can think of a branching function that returns the variable y_j with a free index and having the largest ϕ_j value. Again, in order to save execution time, we consider the following branching function that generates the ordering of indices once for the initial solution and uses it for all subproblems.

Branching Function 3 *Define the fitness ϕ_i of the index i as*

$$\phi_i = \begin{cases} \max\{-a_i, t_i + a_i\}. & \text{if } i \text{ is free in the initial solution,} \\ -\infty & \text{otherwise.} \end{cases}$$

Generate an ordering of indices $1, \ldots, m$, such that index p precedes index q only if $\phi_p \leq \phi_q$. Return the variable with the minimum free index in this ordering.

In the remainder of this section we will assume the existence of a function *FindBranchingVariable* that takes a partial solution as input, and returns the best variable to branch on.

```
function BnP (y: Partial Solution)
begin
    if y is a complete solution
    begin
        update best if necessary;
        return;
    end;
    y ← PegPartialSolution(y);
    y_k ← FindBranchingVariable(y);
    set y_k ← 0;
    if B(y) < z(best) then BnP(y);
    set y_k ← 1;
    if B(y) < z(best) then BnP(y);
    return;
end;
```

Note:
best: Best solution found so far;
$z(\cdot)$: A function to compute the cost of a solution.
$B(\cdot)$: A function to compute the bound from a partial solution.

Fig. 4. Pseudocodes for BnP algorithms

The pseudocode for a recursive implementation of BnP algorithms are presented in Figure 4. We implemented BnB and BnP algorithms to evaluate their performance on randomly generated problem instances as well as on benchmark problem instances. The BnB algorithm was implemented using Branching Function 1. The BnP algorithms were implemented using each of the three branching functions. Notice that we use preprocessing, (using the *PegPartialSolution* function) for both BnB and BnP algorithms. The pseudocode for the bound used in all the implementations is presented in Figure 5. It is an adaptation of a similar bound for general supermodular functions for the SPLP. The algorithms were implemented to allow a maximum execution time of 600 CPU seconds per SPLP instance. The codes were written in C, and run on a Pentium 200 MHz computer running Redhat Linux.

```
function B (y: Partial Solution)
begin
    S = {j : y_j = 0};
    T = {j : y_j ≠ 1};
    l_1 = z(S) − Σ_{k∈T\S} max{0, z(S) − z(S + k)};
    l_2 = z(T) − Σ_{k∈T\S} max{0, z(T) − z(T − k)};
    return max{l_1, l_2};
end;
```

Note: $z(P)$ is assumed to compute the cost of a solution $\mathbf{y}$ such that $y_k = 0 \iff k \in P$.

Fig. 5. Pseudocode for the bound used in the implementations

4   Computational Experiments

We have tested the BnB and BnP algorithms on two types of instances; randomly generated, and benchmarks from the OR-Library ([ ]). The random problem instances were generated in sets of 10 instances each. A problem set is identified by three parameters — the cardinality of the set I (i.e., m), that of the set J (i.e., n), and the density index γ. γ indicates the probability with which an element in the cost matrix has a finite value. Care is taken that while generating the instances, regardless of the γ value, each client can be supplied from a plant in at least one of the candidate sites at finite cost. In each of the randomly generated instances, the fixed costs were chosen from a uniform distribution supported on $[10.0, 1000.0]$, and the finite transportation costs were chosen from a uniform distribution supported on $[1.0, 100.0]$. The benchmark instances were obtained from the OR-Library [ ]. There are twelve SPLP problem instances in this library, four with $m = 16$ and $n = 50$, four with $m = 25$ and $n = 50$, and four with $m = n = 50$. The density index of the transportation cost matrices for all these instances was $\gamma = 1.0$.

Table 1. Number of instances in each set solved within 600 CPU seconds

m n	γ	BnB			BnP Branching Function
					1 2 3
30 50	0.25	10	10 10		10
	0.50	10	10 10		10
	0.75	10	10 10		10
	1.00	10	10 10		10
40 50	0.25	6	6 6		10
	0.50	10	10 10		10
	0.75	10	10 10		10
	1.00	10	10 10		10
50 50	0.25	1	2 2		8
	0.50	4	7 7		10
	0.75	10	10 10		10
	1.00	10	10 10		10

Table 2. The average number of subproblems generated by the algorithms

m n	γ	Number of common instances	BnB	BnP Branching Function		
				1	2	3
30 50	0.25	10	24330.4	13700.4	13463.4	5573.0
	0.50	10	12769.6	6859.0	6859.0	4448.4
	0.75	10	5426.7	2014.8	1969.6	2635.9
	1.00	10	3301.5	326.7	211.1	203.5
40 50	0.25	6	104624.8	59593.7	53771.3	12887.5
	0.50	10	51927.5	26103.9	26103.9	12218.2
	0.75	10	15420.8	5829.7	5829.7	6400.5
	1.00	10	8799.5	806.6	481.8	498.3
50 50	0.25	*				
	0.50	4	62991.25	29188.25	29188.25	17732.25
	0.75	10	37898.2	14327.6	14327.6	13043.5
	1.00	10	19266.2	1391.9	766.5	932.0

* There were too few instances in common.

Table 3. The average execution times required by the algorithms

m n	γ	Number of common instances	BnB	BnP Branching Function		
				1	2	3
30 50	0.25	10	34.190	27.485	26.843	11.074
	0.50	10	20.515	15.252	15.145	9.252
	0.75	10	8.194	4.966	4.843	5.371
	1.00	10	4.355	0.916	0.698	0.742
40 50	0.25	6	240.357	186.698	164.843	41.708
	0.50	10	131.790	58.297	90.792	40.553
	0.75	10	38.862	22.656	22.407	21.126
	1.00	10	19.482	3.184	2.189	2.535
50 50	0.25	*				
	0.50	4	225.255	152.065	150.498	92.313
	0.75	10	139.090	76.566	75.898	62.987
	1.00	10	62.634	7.626	4.781	6.417

* There were too few instances in common.

Table 4. Computational experience with the instances in the OR-Library

Instance	m	n	Number of Subproblems				Execution Times			
			BnB	BnP			BnB	BnP		
				Branching Function				Branching Function		
				1	2	3		1	2	3
cap71	30	50	24	18	19	14	<0.01	<0.01	0.01	0.01
cap72	30	50	37	18	21	13	<0.01	<0.01	0.01	0.01
cap73	30	50	194	130	63	65	0.08	0.07	0.03	0.03
cap74	30	50	63	55	11	37	0.01	0.01	0.01	0.01
cap101	40	50	151	92	92	100	0.05	0.04	0.06	0.07
cap102	40	50	567	325	138	965	0.37	0.31	0.12	0.79
cap103	40	50	2054	589	71	198	1.54	0.62	0.09	0.24
cap104	40	50	943	268	38	72	0.74	0.27	0.09	0.08
cap131	50	50	92543	14148	2167	8016	189.88	35.99	6.24	29.61
cap132	50	50	58564	11234	1226	6992	96.82	22.93	3.29	17.94
cap133	50	50	57697	6459	503	1937	116.88	16.92	1.58	5.92
cap134	50	50	4134	744	125	307	7.57	1.84	0.43	1.05

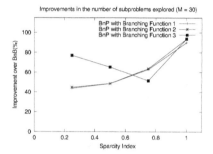
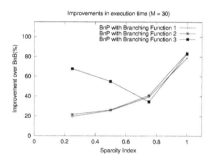

Fig. 6. Performance of BnP algorithms using BnB algorithm as a basis

Tables 1 to 4 present the results of our computations. Table 1 shows the number of problem instances in each data set that were solved by the various algorithms within the stipulated time. Tables 2 and 3 make a comparative study of the average number of subproblems generated by each of the algorithms and the average execution times, based on the instances in the set that were solved by all the algorithms within the stipulated time. Table 4 summarizes our computational experience with the benchmark instances in the OR-Library, presenting both the number of subproblems generated and the execution times required by the algorithms.

The tables show that BnP algorithms in general perform much better than BnB algorithms using the same combinatorial bound. They generate less than 60% of the number of subproblems, and require less than 80% of the execution time for instances with sparse transportation cost matrices. For dense transportation cost matrices, the performance of BnP algorithms is much better — they generate less than 10% of the number of subproblems, and require less than 10% of the execution time. The relative performance of these algorithms improve slightly as the size of the instances increase. The BnB algorithm and BnP algorithms using Branching Functions 1 and 2 find instances with low values of γ more difficult to solve since in these problems, the number of sites where plants are located is close to $\frac{m}{2}$. However BnP algorithms using Branching Function 3 solve these instances efficiently. Figure 6 presents the improvements by the BnP algorithms over BnB algorithms, both in terms of the number of subproblems generated and in terms of the execution times. The shapes of the component graphs do not change for problem instances of larger size. Based on these observations we can conclude that it is better to run a BnP algorithm that uses Branching Function 2 if we know that the transportation matrix is dense (i.e. $\gamma \gtrsim 0.6$) and to run a BnP algorithm that uses Branching Function 3 otherwise. This strategy is verified from the results on the instances in the OR-Library. They have dense transportation cost matrices ($\gamma = 1.0$) and BnP algorithms with Branching Function 2 outperform other algorithms for all instances except cap101 (in which BnP with Branching Function 1 outperforms the rest).

5   Summary and Future Research Directions

In this paper we present branch and peg algorithms for the simple plant location problem (SPLP). These algorithms make two improvements on the basic branch and bound scheme. First, for each subproblem generated in the branch and bound tree, a powerful pegging procedure is applied to reduce the size of the subproblem. Second, the branching function is based on predictions made using the Beresnev function of the subproblem at hand. We see that branch and peg algorithms comprehensively outperform branch and bound algorithms using the same bound, taking less than 10% of the execution time of branch and bound algorithms when the transportation cost matrix is dense. We demonstrate how the coefficients of the linear terms in the Beresnev function play a crucial role in reducing the size of the current subproblem (Pegging Rule), and allow us to predict the potential aggregation of linear and quadratic terms by pegging a variable. This is used in the design of different branching functions. If the transportation cost matrix is sufficiently dense, then based on our computational experience, we recommend a branching function based on a look-ahead scheme, that computes the sizes of the subproblems generated by pegging each variable in the current partial solution, and returns the variable that yields the subproblem of smallest size, as the branching variable (Branching Rule 2). Otherwise, we recommend a branching rule that predicts the variable that is most likely to remain free in all subproblems of the current one, and returns it as a branching variable (Branching Function 3).

The algorithms developed and tested in this paper employ a depth first search scheme. This scheme uses very little computer memory for its execution. However best first search schemes are more useful if we want to generate the minimum number of subproblems. The pegging rule and the branching functions developed in this paper can easily be implemented for branch and bound algorithms using depth first search schemes. It may be interesting to perform computational experiments on branch and peg algorithms using best first search. It may also be interesting to see how the two algorithms compare when other bounds are used.

Branching Functions 2 and 3 described in Section 3 need to compute the ordering of the indices only once. This makes them very time-efficient. But the implicit assumption that this ordering of indices is effective for *all* subproblems in terms of the effectiveness of branching, is not true in general. For example, a modification of Branching Function 2 that calculates the ordering of indices for branching based on the *current* subproblem is much more efficient than the original in terms of the number of subproblems generated. However, the time required to compute the order is prohibitive, which makes this branching function impractical for all but very small instances. An interesting direction of research is to develop book-keeping techniques that would speed up the calculation of the ordering of indices and make such branching functions competitive. Also, we have only used the coefficients of the linear terms in the Beresnev function to develop branching functions. It may be interesting to see whether the non-linear terms in the Beresnev function can give rise to more effective branching strategies.

References

1. J. E. Beasley. OR-Library, http://mscmga.ms.ic.ac.uk/info.html 113
2. V. L. Beresnev. On a Problem of Mathematical Standardization Theory. Upravli-ajemyje Sistemy 11, 43–54, 1973 (in Russian). 107
3. V. L. Beresnev, E. Kh. Gimadi, V. T. Dementyev. Extremal Standardization Problems, Novosibirsk, Nauka, 1978 (in Russian). 106, 107
4. N. Christofides. Graph Theory: An Algorithmic Approach. Academic Press Inc. Ltd., London, 1975. 106
5. G. Cornuejols, G. L. Nemhauser, and L. A. Wolsey. The Uncapacitated Facility Location Problem. Ch.3, Discrete Location Theory, R. L. Francis and P. B. Mirchandani (eds.), Wiley-Interscience, New York, 1990. 106, 107
6. P. M. Dearing, P. L. Hammer, B. Simeone, Boolean and Graph Theoretic Formulations of the Simple Plant Location Problem. Transportation Science 26, 138–148, 1992. 107
7. B. Goldengorin. Requirements of Standards: Optimization Models and Algorithms. ROR, Hoogezand, The Netherlands, 1995. 106
8. B. Goldengorin, D. Ghosh, and G.Sierksma. Equivalent Instances of the Simple Plant Location Problem. SOM Research Report-00A54, 2000. 107, 108, 109
9. P. L. Hammer. Plant Location — A Pseudo-Boolean Approach. Israel Journal of Technology 6, 330–332, 1968. 107
10. P. C. Jones, T. J. Lowe, G. Muller, N. Xu, Y. Ye and J. L. Zydiak. Specially Structured Uncapacitated Facility Location Problems. Operations Research 43, 661–669, 1995. 106
11. D. W. Pentico. The Discrete Two-Dimensional Assortment Problem. Operations Research 36, 324–332, 1988. 106
12. C. S. Revelle and G. Laporte. The Plant Location Problem: New Models and Research Prospects. Operations Research 44, 864–874, 1996. 106
13. A. Tripathy, Süral, and Y. Gerchak. Multidimensional Assortment Problem with an Application. Networks 33, 239–245, 1999. 106

Exploiting Partial Knowledge of Satisfying Assignments

Kazuo Iwama and Suguru Tamaki

School of Informatics, Kyoto University
Kyoto 606-8501, Japan
{iwama,tamak}@kuis.kyoto-u.ac.jp

Abstract. Recently Schöning has shown that a simple local-search algorithm for 3SAT achieves the currently best upper bound, i.e., an expected time of 1.334^n. In this paper, we show that this algorithm can be modified to run much faster if there is some kind of imbalance in satisfying assignments and we have a (partial) knowledge about that. Especially if a satisfying assignment has imbalanced 0's and 1's, i.e., $p_1 n$ 1's and $(1 - p_1)n$ 0's, then we can find a solution in time 1.260^n when $p_1 = 1/3$ and 1.072^n when $p_1 = 0.1$. Such an imbalance often exists in SAT instances reduced from other combinatorial problems. As a concrete example, we investigate a reduction from 3DM and show our new approach is nontrivially faster than its direct algorithms. Preliminary experimental results are also given.

1 Introduction

In [16], Schöning gave the celebrated randomized algorithm for the CNF Satisfiability Problem (SAT), which runs in an expected time of 1.334^n (multiplied by a polynomial). The algorithm is based on simple local search, i.e., (i) selecting an initial assignment at random, (ii) selecting an arbitrary unsatisfied clause and flipping one of the variables in it, and (iii) repeat (ii) $3n$ times. He proved that the possibility p of successfully finding a satisfying assignment by this procedure is

$$p \geq \left(\frac{1}{2} \left(1 + \frac{1}{k-1} \right) \right)^n , \tag{1}$$

where k is the maximum number of literals in each clause. In the case of 3SAT, the value of the right hand size is $(3/4)^n$. In other words, we can find a satisfying assignment with high probability by repeating the above procedure roughly $(4/3)^n$ times (multiplied by a polynomial).

In this paper, we first give a generalization of the equation (1), namely, we prove that

$$p \geq \prod_{i=1}^{n} \left(t_i + \frac{f_i}{k-1} \right) , \tag{2}$$

G. Brodal et al. (Eds.): WAE 2001, LNCS 2141, pp. 118–128, 2001.
© Springer-Verlag Berlin Heidelberg 2001

where t_i ($f_i = 1 - t_i$, resp.) is the probability that the variable x_i is assigned a correct (incorrect, resp.) value at the initialization step. (If $t_i = f_i = 1/2$, then (2) is the same as (1).) This equation says that if we have some knowledge on the value of x_i in a satisfying assignment, we can increase the success probability. For example, suppose that we know for some reason, 90% of the odd-indexed variables $x_1, x_3, x_5 \ldots$ take value 1 in a satisfying assignment. Then our best strategy is to select 1 initially for all the odd-indexed variables and to select 0 or 1 at random for the even-indexed variables. Then the success probability (when $k = 3$) calculated from (2) is

$$p \geq 1^{0.45n} \left(\frac{1}{2}\right)^{0.05n} \left(\frac{3}{4}\right)^{0.5n} \cong 0.836^n.$$

This means that we would be able to obtain solution in roughly $(1/0.836)^n = 1.196^n$ steps , which is much better than the original 1.334^n.

As a concrete example of such a partial knowledge of solutions, we consider an imbalance between the number of 0's and 1's in the satisfying assignment. Suppose that we know the satisfying assignment includes $p_0 n$ 0's and $p_1 n$ 1's ($0 \leq p_0 \leq 1$ and $p_1 = 1 - p_0$). Then we can obtain optimal probabilities q_0 and q_1 ($= 1 - q_0$) by using (2), such that we should assign 0 to each variable with probability q_0 and 1 with q_1 at the beginning. Our result shows that the expected time complexity when we use this optimal initial-assignment is

$$T = \begin{cases} \left(\frac{1}{p_0}\right)^{p_0 n} \left(\frac{1}{p_1}\right)^{p_1 n} \left(\frac{k-1}{k}\right)^n & \text{for } \frac{1}{k} \leq p_0 \leq \frac{k-1}{k}, \\ (k-1)^{\min\{p_0 n, p_1 n\}} & \text{for } p_0 < \frac{1}{k} \text{ or } p_0 > \frac{k-1}{k}. \end{cases}$$

For example, when $p_0 = 2/3$, $T = 1.260^n$ and when $p_0 = 0.9$, $T = 1.072^n$. Such an imbalance of 0's and 1's often appears in instances encoded from other problems. For example, SAT-instances encoded from the class-schedule problem [,] have the property that solutions must have very few 1's. Also, let us remember the famous result by Cook [ ] where SAT is first proved to be NP-complete. One can see that his reduction also has the same property.

In this paper, we take a more combinatorial problem, i.e., 3-Dimensional Matching (3DM), as a concrete example of such an imbalance. An instance of 3DM is given as (W, X, Y, M) where $|W| = |X| = |Y| = q$ and $M \subseteq W \times X \times Y$. If each element in $W \cup X \cup Y$ appears in M evenly, i.e., all k (or less) times, then our reduction gives a kSAT instance using kq variables. Our reduction also assures that any satisfying assignment has exactly q 1's against the kq variables. In other words, the resulting formulas do have the imbalance whose degree is represented by $p_1 = 1/k$. Note that this reduction is quite natural and it appears hard to come up with another reduction (whether or not it creates the imbalance) which provides reasonably simple formulas.

We also show experimental results although they are preliminary. Our instances are those encoded from 3DM and from prime factorization. It is clearly demonstrated that our approach is faster than the original Schöning, especially for the second set of instances. Note that the second instances are harder than the other since the number of satisfying assignments is few.

It is needless to say that satisfiability testing has been one of the most popular research topics in both theoretical and practical computer science. Even if we focus on the worst-case analyses (deterministic or randomized), there have been many papers such as [5,7,8,10,11], where the Schöning's algorithm achieves the best upper bound. (Very recently a slight improvement was reported in [14].) Our result is not a general improvement, but we believe that there are many cases for which our approach is useful.

2   Schöning's Algorithm

In this paper, a clause is denoted like $(x_1 \vee \neg x_2 \vee x_3)$ and a CNF-formula is a conjunction of clauses such as $f = (x_1 \vee \neg x_2 \vee x_3) \wedge (x_3 \vee \neg x_4 \neg \vee x_5) \wedge (x_1 \vee x_3 \vee x_5)$. If each clause in f has at most k literals, f is called a kCNF-formula. kSAT is usually defined as a decision problem, i.e., to answer whether or not a given kCNF-formula has a satisfying assignment. However, since this paper deals with only randomized local search, we assume that all instances given are satisfiable or have at least one satisfying assignment.

Schöning's algorithm, denoted by A_s, is a typical local-search algorithm described as follows: (i) For a given formula f of n variables, select an initial assignment a at random. (ii) If a is a satisfying assignment, then stop. Otherwise, select an arbitrary unsatisfied clause C, select one variable x in C at random, and obtain a new assignment a' by flipping (0 to 1 or 1 to 0) the value of x. Let $a = a'$. (iii) Repeat (ii) $3n$ times. The main result of [15] is given as the following lemma:

Lemma 1 ([15]). *Suppose that f is a kCNF-formula and that A_s starts from an initial assignment a such that the Hamming distance between a and some satisfying assignment a^* of f is j. Then the probability q_j that A_s successfully finds a satisfying assignment is*

$$q_j \geq \left(\frac{1}{k-1} \right)^j.$$

Now one can see that the probability p that A_s finds a satisfying assignment can be written as

$$p \geq \left(\frac{1}{2} \right)^n \sum_{j=0}^{n} \binom{n}{j} \left(\frac{1}{k-1} \right)^j$$

$$= \left(\frac{1}{2} \left(1 + \frac{1}{k-1} \right) \right)^n.$$

It then follows that if we simply repeat A_s $poly(n) \cdot \frac{1}{p} = poly(n) \left(2 - \frac{2}{k} \right)^n$ times, then we can obtain a solution with high probability. In this paper, we omit the term $poly(n)$ and say that A_s runs in time $\left(2 - \frac{2}{k}\right)^n$. When $k = 3$, this time complexity is equal to $\left(\frac{4}{3}\right)^n \cong 1.334^n$.

3   Selection of Initial Assignments

In this section, we first generalize Lemma 1, which is then used to derive improved bounds for kCNF-formulas having the imbalance in their solutions. From now on, when we say algorithm A_s, it means only steps (ii) and (iii); step (i) where an initial assignment is chosen is given explicitly.

Lemma 2. *For some satisfying assignment a^*, let t_i $(1 \leq i \leq n)$ be the probability that variable x_i receives the same (correct) initial assignment as a^*. Also let $f_i = 1 - t_i$. Then the probability p that A_s is successful is*

$$p \geq \prod_{i=1}^{n} \left(t_i + \frac{f_i}{k-1}\right).$$

Proof. Suppose that X (X', resp.) is a random variable such that X (X', resp.) variables among $x_1, \ldots, x_n$ (among $x_2, \ldots, x_n$, resp.) receive incorrect values in the initial assignment. Then by Lemma 1, the probability p can be written as

$$p \geq \sum_{j=0}^{n} Pr\{X = j\} \left(\frac{1}{k-1}\right)^j$$

$$= t_1 \sum_{j=0}^{n-1} Pr\{X' = j\} \left(\frac{1}{k-1}\right)^j + f_1 \sum_{j=0}^{n-1} Pr\{X' = j\} \left(\frac{1}{k-1}\right)^{j+1}$$

$$= \left(t_1 + \frac{f_1}{k-1}\right) \sum_{j=0}^{n-1} Pr\{X' = j\} \left(\frac{1}{k-1}\right)^j$$

Applying a similar reduction to the summation term $n - 1$ times, we can get the formula in the lemma.   □

Now we consider kCNF-formulas having the imbalance in their solutions. Suppose that a given formula f has a satisfying assignment a^* which has l 0's and $n - l$ 1's. Let $p_0 = l/n$ and $p_1 = (n - l)/n$. Our new algorithm, denoted by $A_s(p_0)$, differs from A_s only in step (i): Namely, each variable x_i is assigned 0 with probability q_0 and is assigned 1 with probability q_1, where the value of q_0 is given by the following theorem.

Theorem 1. *Let p be the probability that Algorithm $A_s(p_0)$ is successful. Then p becomes maximum when the probability q_0 with which each variable is assigned 0 initially is given as*

$$q_0 = \begin{cases} 1 & \text{for } p_0 < \frac{1}{k}, \\ \dfrac{kp_0 - 1}{k-2} & \text{for } \frac{1}{k} \leq p_0 \leq \frac{k-1}{k}, \\ 0 & \text{for } p_0 > \frac{k-1}{k}, \end{cases}$$

and the value of p for this optimal q_0 is

$$p \geq \begin{cases} p_0^{p_0 n} p_1^{p_1 n} \left(\dfrac{k}{k-1}\right)^n & \text{for } \dfrac{1}{k} \leq p_0 \leq \dfrac{k-1}{k}, \\ \left(\dfrac{1}{k-1}\right)^{\min\{p_0 n,\, p_1 n\}} & \text{for } p_0 < \dfrac{1}{k} \text{ or } p_0 > \dfrac{k-1}{k}. \end{cases}$$

Proof. By Lemma 2, the probability p can be written as

$$p \geq \prod_{i=0}^{n} \left(t_i + \frac{f_i}{k-1}\right) = \left(q_0 + \frac{q_1}{k-1}\right)^{p_0 n} \left(q_1 + \frac{q_0}{k-1}\right)^{p_1 n}.$$

To decide the value of q_0 that maximizes p, we consider the following function

$$\sigma(q_0) = \log \left\{ \left(q_0 + \frac{q_1}{k-1}\right)^{p_0} \left(q_1 + \frac{q_0}{k-1}\right)^{p_1} \right\}$$

$$= p_0 \log \left\{ \left(1 - \frac{1}{k-1}\right) q_0 + \frac{1}{k-1} \right\} + (1 - p_0) \log \left\{ -\left(1 - \frac{1}{k-1}\right) q_0 + 1 \right\}.$$

σ is convex in $[0, 1]$, so it takes maximum value where its derivative is 0 or at either end of the interval $[0, 1]$. Since

$$\sigma'(q_0) = p_0 \frac{\left(1 - \frac{1}{k-1}\right)}{\left(1 - \frac{1}{k-1}\right) q_0 + \frac{1}{k-1}} + (1 - p_0) \frac{-\left(1 - \frac{1}{k-1}\right)}{-\left(1 - \frac{1}{k-1}\right) q_0 + 1},$$

$\sigma'(q_0) = 0$ implies $q_0 = (kp_0 - 1)/(k - 2)$. Substituting this optimal q_0, or substituting $q_0 = 0$ or $q_0 = 1$ if $(kp_0 - 1)/(k - 2)$ is less than 0 or greater than 1, respectively, we obtain the theorem. □

Remark *The value of q_0 is quite different from the value of p_0. For example, if $p_0 = 0.6$ and $k = 3$, the value of q_0 is 0.8, and if $p_0 \geq 2/3$, then $q_0 = 1.0$. Namely, the imbalance should be expanded in the initial assignment.*

Fig. 1 shows numerical examples of Theorem 1 for $k = 3, 4, 5$, and 6. The horizontal axis shows the value of $0 \leq p_0 \leq 1$ and the vertical axis shows the value of c supposing that the optimal bound of Theorem 1 is represented as c^n. Note that the time complexity is roughly bounded by $1/p$.

4   Three Dimensional Matching

An instance of 3DM is given as (W, X, Y, M) where W, X and Y are disjoint sets of size q and M ($|M| = n$) is a subset of $W \times X \times Y$. Its question is whether or not there is a subset $M' \subseteq M$ such that $|M'| = q$ and all elements

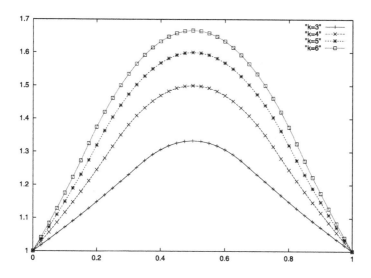

Fig. 1. Numerical examples of Theorem 1

in $W \times X \times Y$ appear (exactly once) in M'. For an integer k, k3DM is a restricted version of 3DM, namely, each element in $W \times X \times Y$ appears at most k times in M (and therefore $n \leq kq$). k3DM can be reduced to kSAT as follows: For given (W, X, Y, M), we construct a formula F such that: (i) F uses n variables $z_1, z_2, \ldots, z_n$ (z_i corresponds to the ith triple in M). (ii) Suppose that an element $w \in W$ appears in the i_1th, i_2th, $\ldots$, i_kth triples in M. Then we prepare a CNF-formula $U^w(z_{i_1}, z_{i_2}, \ldots, z_{i_k})$ such that it becomes 1 if and only if exactly one of $z_{i_1}, z_{i_2}, \ldots, z_{i_k}$ is 1. When $k = 3$, for example, $U^w(z_{i_1}, z_{i_2}, z_{i_3})$ can be written as

$$(\neg z_{i_1} \vee \neg z_{i_2} \vee \neg z_{i_3}) \wedge (\neg z_{i_1} \vee \neg z_{i_2} \vee z_{i_3}) \wedge (z_{i_1} \vee \neg z_{i_2} \vee \neg z_{i_3})$$
$$\wedge (\neg z_{i_1} \vee z_{i_2} \vee \neg z_{i_3}) \wedge (z_{i_1} \vee z_{i_2} \vee z_{i_3}).$$

(iii) The entire formula F is a conjunction of U^w for all $w \in W$, U^x for all $x \in X$ and U^y for all $y \in Y$.

One can see easily that (i) F is satisfiable iff the original (W, X, Y, M) has a matching, and (ii) if F is satisfiable, then any solution has q 1's, i.e., an imbalanced satisfying assignment. For example, if $k = 3$, then $p_1 = 1/3$ and $A_s(2/3)$ finds a solution in time 1.260^n and if $k = 4$, then $A_s(1/4)$ for 4SAT dose so in time 1.317^n.

For a comparison purpose, let us consider a naive method of solving k3DM directly. Since each element in W appears in M at most k times, there are at most k^q different ways of selecting q (or less) triples from M which cover all elements in W. One can compute whether or not these q triples constitute a matching in polynomial time. Thus the time complexity of this algorithm can be written as $k^{n/k}$. This is 1.443^n for $k = 3$ and 1.588^n for $k = 4$. In both cases our bounds of $A_s(1/k)$ are much better. One can see that it is hard to find other reductions which are reasonably simple, whether or not their satisfying assignments are balanced.

5 Experiments

Experiments were conducted for CNF-formulas reduced from 3DM and from prime factorization. For the 3DM formulas, we first obtain a random 3DM instance by generating n triples which are to be in M. This generation is basically random but (i) to assure that the instance has a matching, we first generate an artificial matching (of q triples) and then (ii) add $n - q$ triples so that each element in $W \times X \times Y$ appears exactly three times. This 3DM instance is reduced to a 3SAT instance as described in Section 4.

We have generated 15 different formulas for each of $n = 250, 500, 750, 1000$ and 1250. We tested six different local-search algorithms:

(1) Pure GSAT [ ].
(2) Weighting [, , ,].
(3) Weighting + 50% Random Walk.
(4) GSAT + 50% Random Walk [ ].
(5) Schöning.
(6) Our modified Schöning.

Each algorithm is run 100 times (using different random numbers) for each single instance. Since (5) and (6) execute $3n$ flips in a single try, we also execute the same number of flips in other algorithms. (Since each algorithm has its own recommended value for the number of flips in a single try, this setting might not be too fair.) The result is given in Table 1, which shows the ratio of successful tries (average values over 15 instances). Algorithms (1) through (6) are denoted by g, gw, gwn, gn, A_s and $A_s(2/3)$, respectively. We also tested algorithms g, gw, gwn and gn for the imbalanced initial assignments as $A_s(2/3)$, but we did not find clear differences.

Table 1. 3DM Instances

# of variables	g	gw	gwn	gn	A_s	$A_s(2/3)$
250	0.129	0.176	0.528	0.475	0.468	0.613
500	0.002	0.058	0.4355	0.41	0.405	0.581
750	0.00	0.019	0.368	0.388	0.428	0.599
1000	0.00	0.015	0.37	0.385	0.357	0.585
1250	0.00	0.007	0.345	0.354	0.367	0.631

One can immediately see that the absolute success ratio is very high compared to the analysis. Obvious reason is that each instance has a lot of satisfying assignments; there is a good chance that one of them happens to be quite close to the initial assignment chosen by the algorithm. This nature certainly discourages the effort of selecting initial assignments cleverly. Even so, our $A_s(2/3)$ is clearly better than others.

Table 2. Effect of the value of q_0

q_0	0.5	0.6	0.7	0.8	0.9	1.0
	0.356	0.475	0.576	0.634	0.648	0.598

Note that $A_s(2/3)$ initially assigns 0 to all the variables (i.e., $q_0 = 1.0$) by Theorem 1. Table 2 shows how the success ratio changes with the value of q_0 by using five 3DM formulas of 1250 variables. Our algorithm is run 5000 times for each instance for $q_0 = 0.5, 0.6, 0.7, 0.8, 0.9$ and 1.0 and the table shows the average success ratio of five instances. As mentioned above, $A_s(2/3)$ becomes optimal for $q_0 = 1.0$. However, the experiments suggest the optimal point exists around $q_0 = 0.9$ probably for the same reason mentioned before.

Another benchmark is a reduction from prime factorization. Again instances are 3CNF-formulas, which are denoted by P_n. If the integer n can be represented by $n = n_1 \times n_2$ for integers n_1 and $n_2 > 1$, then P_n has a single satisfying assignment corresponding to this pair of n_1 and n_2. Hence, if n is a product of two prime numbers, P_n has only one satisfying assignment. P_n is constructed by simulating the usual multiplication procedure using many auxiliary variables other than those used for binary representations of n_1 and n_2. Our experiment has used P_{129} which uses 136 variables and contains 337 clauses. Since $129 = 3 \times 43$, P_{129} has only one satisfying assignment. To make the imbalanced situation, we flip the polarities of the variables appropriately so that the satisfying assignment has 50%, 60%, 70%, 80% and 90% 0's.

Table 3 shows how many times $A_s(q_0)$ succeeds out of 100,000 tries. Each column corresponds to the imbalance described above (50% for the first column and 90% for the last one). Note that the optimal q_0 for the 60% imbalance is 0.8 due to Theorem 1. Note that P_{129} includes a lot of clauses which contains only one or two literals, which obviously makes it easy to solve.

Our third benchmark was taken from the DIMACS benchmark set. We tested only one instance called aim-50-1_6-yes1-1.cnf, which is basically a random 3SAT instance (each clause includes exactly three literals) and has 50 variables and 80 clauses. Also, it has only one satisfying assignment. Thus this instance appears the hardest among what we used in the experiments. Like Table 3, Table 4 shows how many times $A_s(q_0)$ succeeds out of 1,000,000 tries. (Note that the number of tries is five times as many as Table 3.)

Table 3. Prime-factorization instances

p_0	q_0					
	0.5	0.6	0.7	0.8	0.9	1.0
0.5	5	5	5	11	4	8
0.6	4	11	8	14	16	8
0.7	6	6	15	20	28	50
0.8	6	15	28	74	179	520
0.9	9	4	61	235	994	5063

Table 4. DIMACS instances

p_0	q_0					
	0.5	0.6	0.7	0.8	0.9	1.0
0.5	15	9	11	9	9	11
0.6	18	21	20	33	50	48
0.7	16	35	60	134	273	630
0.8	12	38	135	381	1176	2867
0.9	14	55	350	1898	10943	50424

6   Constraint Satisfaction Problem

Schöning shows in [ ] that a similar local search algorithm is also efficient for Constraint Satisfaction Problem (CSP). An instance of CSP is a set of constraints $C_1, C_2, \ldots, C_m$ and each constraint $C_i(x_1, x_2, \ldots, x_n)$ is a function from $\{0, 1, \ldots, d-1\}$ into $\{0, 1\}$. Each variable x_i takes one of the d different values. He proves that the algorithm finds an answer in an expected time of

$$\left\{ d \left(1 - \frac{1}{l} \right) \right\}^n,$$

where l is the degree of the instance, namely, each C_i depends on at most l variables.

Suppose that there is a similar imbalance in a solution, such that $p_0 n$ variables take value 0 (and the other $(1 - p_0)n$ ones take 1 through $d - 1$). Then by changing the initial assignment exactly as before, we can improve the expected time complexity T as follows (proof is omitted).

$$
T = \begin{cases}
\left(\dfrac{d-1}{h-1}\right)^{-p_0 n} \left(1 + \dfrac{d-2}{h-1}\right)^{-p_1 n} \\
\qquad\qquad\qquad \text{for } p_0 < \dfrac{1}{h+d-2}, \\[2ex]
\left(q_0 + \dfrac{(d-1)q_1}{h-1}\right)^{-p_0 n} \left(q_1 + \dfrac{q_0 + (d-2)q_1}{h-1}\right)^{-p_1 n} \\
\qquad\qquad \text{for } \dfrac{1}{h+d-2} \leq p_0 \leq \dfrac{h-1}{h+d-2}, \\[2ex]
\left(\dfrac{d-1}{h-1}\right)^{-p_0 n} \\
\qquad\qquad\qquad \text{for } p_0 > \dfrac{h-1}{h+d-2},
\end{cases}
$$

where $p_1 = 1 - p_0$, $q_0 = \dfrac{(h+d-2)p_0 - 1}{h-2}$, $q_1 = 1 - \dfrac{q_0}{d-1}$
and $h = (l-1)(d-1) + 1$.

7   Concluding Remarks

Lemma 2 is quite useful when we analyze the performance of A_s with a modified initialization step. Its performance deeply depends on the selection of initial assignment, which will continue to be an important research issue. Experimental results in this paper are apparently not enough, which will be strengthened in future reports. The programs for the algorithms (1) to (4) (see Sec.5) have been developed by our research group, mainly, by Daiji Fukagawa.

References

1. B. Cha and K. Iwama. Performance test of local search algorithms using new types of random CNF formulas. *Proc. IJCAI-95*, 304–310, 1995. 124
2. B. Cha and K. Iwama. Adding new clauses for faster local search. *Proc. AAAI'96*, 332-337, 1996. 124
3. B. Cha, K. Iwama, Y. Kambayashi, and S. Miyazaki. Local search algorithms for partial MAXSAT. *Proc. AAAI'97*, 263–268, 1997. 119
4. S. A. Cook. The complexity of theorem-proving procedures. In *Proceedings of the 3rd Annual ACM Symposium on the Theory of Computing*, 151–158, 1971. 119
5. E. Dantsin, A. Goerdt, E. A. Hirsch, R. Kannan, J. Kleinberg, C. Papadimitriou, P. Raghavan, and U. Schöning. A deterministic $2 - \frac{2}{k+1}$ algorithm for k-SAT based on local search. submitted, 2000. 120
6. K. Iwama, D. Kawai, S. Miyazaki, Y. Okabe, and J. Umemoto. Parallelizing local search for CNF satisfiability using vectorization and PVM. *Proc. Workshop on Algorithm Engineering (WAE 2000)*. 119
7. O. Kullmann, New methods for 3-SAT decision and worst-case analysis. *Theoretical Computer Science*, 223(1-2):1–72, 1999. 120

8. B. Monien and E. Speckenmeyer. Solving satisfiability less than 2^n steps. *Discrete Applied Mathematics*, 10:287–295, 1985. 120
9. P. Morris. The breakout method for escaping from local search minima. *Proc. AAAI'93*, 40–45, 1993. 124
10. R. Paturi, P. Pudlák, M. E. Saks, and F. Zane. An improved exponential-time algorithm for k-SAT. *Proceedings 39th Annual Symposium on Foundations of Computer Science*, 628–637, 1998. 120
11. R. Paturi, P. Pudlák, and F. Zane. Satisfiability coding lemma. *Proceedings 38th Annual Symposium on Foundations of Computer Science*, 566–574, 1997. 120
12. B. Selman and H. Kautz. Local search strategy for satisfiability testing. 2nd DIMACS challenge Workshop, 1993. 124
13. B. Selman, H. Levesque, and D. Mitchell. A new method for solving hard satisfiability problems. *Proc. AAAI-92*, 440–446, 1992. 124
14. R. Schuler, U. Schöning, and O. Watanabe. An improved randomized algorithm for 3-SAT. Preprint, 2001. 120
15. U. Schöning. A probabilistic algorithm for k-SAT and constraint satisfaction problems. *Proceedings 40th Annual Symposium on Foundations of Computer Science*, 410-414, 1999. 118, 120, 126

Using PRAM Algorithms on a Uniform-Memory-Access Shared-Memory Architecture

David A. Bader[1], Ajith K. Illendula[2], Bernard M.E. Moret[3], and
Nina R. Weisse-Bernstein[4]

[1] Department of Electrical and Computer Engineering, University of New Mexico
Albuquerque, NM 87131 USA
dbader@eece.unm.edu[‡]
[2] Intel Corporation, Rio Rancho, NM 87124 USA
ajith.k.illendula@intel.com
[3] Department of Computer Science, University of New Mexico
Albuquerque, NM 87131 USA
moret@cs.unm.edu[§]
[4] University of New Mexico, Albuquerque, NM 87131 USA
nina@unm.edu[¶]

Abstract. The ability to provide uniform shared-memory access to a
significant number of processors in a single SMP node brings us much
closer to the ideal PRAM parallel computer. In this paper, we develop
new techniques for designing a uniform shared-memory algorithm from
a PRAM algorithm and present the results of an extensive experimental
study demonstrating that the resulting programs scale nearly linearly
across a significant range of processors (from 1 to 64) and across the
entire range of instance sizes tested. This linear speedup with the number
of processors is, to our knowledge, the first ever attained in practice for
intricate combinatorial problems. The example we present in detail here
is a graph decomposition algorithm that also requires the computation of
a spanning tree; this problem is not only of interest in its own right, but
is representative of a large class of irregular combinatorial problems that
have simple and efficient sequential implementations and fast PRAM
algorithms, but have no known efficient parallel implementations. Our
results thus offer promise for bridging the gap between the theory and
practice of shared-memory parallel algorithms.

1 Introduction

Symmetric multiprocessor (SMP) architectures, in which several processors operate in a true, hardware-based, shared-memory environment and are packaged

[‡] Supported in part by NSF CAREER 00-93039, NSF ITR 00-81404, NSF DEB 99-
10123, and DOE CSRI-14968
[§] Supported in part by NSF ITR 00-81404
[¶] Supported by an NSF Research Experience for Undergraduates (REU)

G. Brodal et al. (Eds.): WAE 2001, LNCS 2141, pp. 129–144, 2001.

as a single machine, are becoming commonplace. Indeed, most of the new high-performance computers are clusters of SMPs, with from 2 to 64 processors per node. The ability to provide uniform-memory-access (UMA) shared-memory for a significant number of processors brings us much closer to the ideal parallel computer envisioned over 20 years ago by theoreticians, the *Parallel Random Access Machine (PRAM)* (see [,]) and thus may enable us at long last to take advantage of 20 years of research in PRAM algorithms for various irregular computations. Moreover, as supercomputers increasingly use SMP clusters, SMP computations will play a significant role in supercomputing. For instance, much attention has been devoted lately to OpenMP [ ], which provides compiler directives and runtime support to reveal algorithmic concurrency and thus take advantage of the SMP architecture; and to mixed-mode programming, which combines message-passing style between cluster nodes (using MPI [ ]) and shared-memory style within each SMP (using OpenMP or POSIX threads [ ]).

While an SMP is a shared-memory architecture, it is by no means the PRAM used in theoretical work—synchronization cannot be taken for granted and the number of processors is far smaller than that assumed in PRAM algorithms. The significant feature of SMPs is that they provide much faster access to their shared-memory than an equivalent message-based architecture. Even the largest SMP to date, the 64-processor Sun Enterprise 10000 (E10K), has a worst-case memory access time of 600ns (from any processor to any location within its 64GB memory); in contrast, the latency for access to the memory of another processor in a distributed-memory architecture is measured in tens of μs. In other words, message-based architectures are two orders of magnitude slower than the largest SMPs in terms of their worst-case memory access times.

The largest SMP architecture to date, the Sun E10K [ ], uses a combination of data crossbar switches, multiple snooping buses, and sophisticated cache handling to achieve UMA across the entire memory. Of course, there remains a large difference between the access time for an element in the local processor cache (around 15ns) and that for an element that must be obtained from memory (at most 600ns)—and that difference increases as the number of processors increases, so that cache-aware implementations are, if anything, even more important on large SMPs than on single workstations. Figure 1 illustrates the memory access behavior of the Sun E10K (right) and its smaller sibling, the E4500 (left), using a single processor to visit each node in a circular array. We chose patterns of access with a fixed stride, in powers of 2 (labelled C, stride), as well as a random access pattern (labelled R). The data clearly show the effect of addressing outside the on-chip cache (the first break, at a problem size of 2^{13} words, or 32KB) and then outside the L2 cache (the second break, at a problem size of 2^{21} words, or 8MB). The uniformity of access times was impressive—standard deviations around our reported means are well below 10 percent. Such architectures make it possible to design algorithms targeted specifically at SMPs.

In this paper, we present promising results for writing efficient implementations of PRAM-based parallel algorithms for UMA shared-memory machines. As an example of our methodology, we look at ear decomposition in a graph

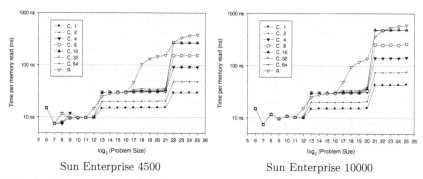

Fig. 1. Memory access (read) time using one MHz processor of a Sun E4500 (left) and an E10K (right) as a function of array size for various strides

and show that our implementation achieves near linear speedups, as illustrated in Figure 2. Our main contributions are

1. A new methodology for designing practical shared-memory algorithms on UMA shared-memory machines,
2. A fast and scalable shared-memory implementation of ear decomposition (including spanning tree construction) demonstrating the first ever significant—in our case, nearly optimal—parallel speedup for this class of problems.
3. An example of experimental performance analysis for a nontrivial parallel implementation.

2 Related Work

Several groups have conducted experimental studies of graph algorithms on parallel architectures [, , , , , ,]. Their approach to producing a parallel program is similar to ours (especially that of Ramachandran *et al.* [ ]), but their test platforms have not provided them with a true, scalable, UMA shared-memory environment or have relied on *ad hoc* hardware [ ]. Thus ours is the first study of speedup over a significant range of processors on a commercially available platform.

3 Methodology

3.1 Approach

Our methodology has two principal components: an approach to the conversion of PRAM algorithms into parallel programs for shared-memory machines and a matching complexity model for the prediction of performance. In addition, we

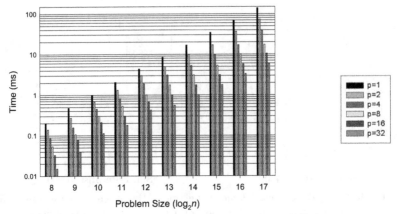

a) varying graph and machine size with fixed random sparse graph.

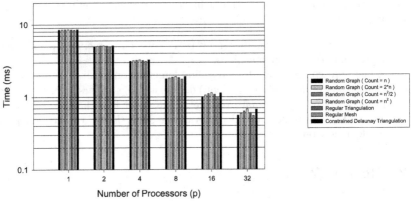

b) varying number of processors and sparse graphs with fixed size $n = 8192$.

Fig. 2. Running times of ear decomposition on the NPACI Sun E10K with 1 to 32 processors a) on varying problem sizes (top) and b) different sparse graph models with $n = 8192$ (bottom)

make use of the best precepts of algorithm engineering [ ] to ensure that our implementations are as efficient as possible.

Converting a PRAM algorithm to a parallel program requires us to address three problems: (i) how to partition the tasks (and data) among the very limited number of processors available; (ii) how to optimize locally as well as globally the use of caches; and (iii) how to minimize the work spent in synchronization (barrier calls). The first two are closely related: good data and task partitioning will ensure good locality; coupling such partitioning with cache-sensitive coding (see [, , ,] for discussions) provides programs that take best advantage of the architecture. Minimizing the work done in synchronization barriers is a fairly simple exercise in program analysis, but turns out to be far more difficult in practice: a tree-based gather-and-broadcast barrier, for instance, will guarantee synchronization of processes at fairly minimal cost (and can often be split when only one processor should remain active), but may not properly synchronize the caches of the various processors on all architectures, while a more onerous barrier that forces the processors' caches to be flushed and resynchronized will be portable across all architectures, but unnecessarily expensive on most. We solve the problem by providing both a simple tree-based barrier and a heavy-weight barrier and placing in our libraries architecture-specific information that can replace the heavy barrier with the light-weight one whenever the architecture permits it.

3.2 Complexity Model for Shared-Memory

Various cost models have been proposed for SMPs [, , , , , , , ,]; we chose the Helman and JáJá model [ ] because it gave us the best match between our analyses and our experimental results. Since the number of processors used in our experiments is relatively small (not exceeding 64), contention at the memory location is negligible compared to the contention at the processors. Other models that take into account only the number of reads/writes by a processor or the contention at the memory are thus unsuitable here. Performance on most computer systems dictates the reliance on several levels of memory caches, and thus, cost benefit should be given to an algorithm that optimizes for contiguous memory accesses over non-contiguous access patterns. Distinguishing between contiguous and non-contiguous memory accesses is the first step towards capturing the effects of the memory hierarchy, since contiguous memory accesses are much more likely to be cache-friendly. The Queuing Shared Memory [,] model takes into account both the number of memory accesses and contention at the memory, but does not distinguish between between contiguous versus non-contiguous accesses. In contrast, the complexity model of Helman and JáJá [ ] takes into account contention at both the processors and the memory. In the Helman-JáJá model, we measure the overall complexity of an algorithm by the triplet (M_A, M_E, T_C), where M_A is the maximum number of (noncontiguous) accesses made by any processor to main memory, M_E is the maximum amount of data exchanged by any processor with main memory, and T_C is an upper bound on the local computational work of any of the processors. T_C is measured in standard asymptotic terms, while M_A and M_E are represented as approximations of the actual values. In practice, it is often possible to focus on either M_A or M_E when examining the cost of algorithms. Because the number of required barrier synchronizations is less than the local computational work on each processor, the cost of synchronization is dominated by the term T_C and thus is not explicitly included in the model.

4   Example: Ear Decomposition

4.1   The Problem

Of the many basic PRAM algorithms we have implemented and tested, we chose ear decomposition to present in this study, for three reasons. First, although the speedup observed with our implementation of ear decomposition is no better than what we observed with our other implementations of basic PRAM algorithms, ear decomposition is more complex than such problems as prefix sum, pointer doubling, symmetry breaking, etc. Secondly, ear decomposition is typical of problems that have simple and efficient sequential solutions, have known fast or optimal PRAM algorithms, but yet have no practical parallel implementation. One of its component tasks is finding a spanning tree, a task that was also part of the original DIMACS challenge on parallel implementation many years ago, in which sequential implementations had proved significantly faster than even massively parallel implementations (using a 65,536-processor CM2). Finally, ear decomposition is interesting in its own right, as it is used in a variety of applications from computational geometry [7, 10, 11, 24, 25], structural engineering [12, 13], to material physics and molecular biology [14].

The efficient parallel solution of many computational problems often requires approaches that depart completely from those used for sequential solutions. In the area of graph algorithms, for instance, depth-first search is the basis of many efficient algorithms, but no efficient PRAM algorithm is known for depth-first search (which is P-complete). To compensate for the lack of such methods, the technique of *ear decomposition*, which does have an efficient PRAM algorithm, is often used [25].

Let $G = (V, E)$ be a connected, undirected graph; set $n = |V|$ and $m = |E|$. An *ear decomposition* of G is a partition of E into an ordered collection of simple paths (called *ears*), $Q_0, Q_1, \ldots, Q_r$, obeying the following properties:

- Each endpoint of Q_i, $i > 0$, is contained in some Q_j, $j < i$.
- No internal vertex of Q_i, $(i > 0)$, is contained in any Q_j, for $j < i$.

Thus a vertex may belong to more than on ear, but an edge is contained in exactly one ear [2]. If the endpoints of the ear do not coincide, then the ear is *open*; otherwise, the ear is *closed*. An *open ear decomposition* is an ear decomposition in which every ear is open. Figure 5 in the appendix illustrates these concepts. Whitney first studied open ear decomposition and showed that a graph has an open ear decomposition if and only if it is biconnected [40]. Lovász showed that the problem of computing an open ear decomposition in parallel is in NC [30]. Ear decomposition has also been used in designing efficient sequential and parallel algorithms for triconnectivity [37] and 4-connectivity [23]. In addition to graph connectivity, ear decomposition has been used in graph embeddings (see [1]).

The sequential algorithm: Ramachandran [37] gave a linear-time algorithm for ear decomposition based on depth-first search. Another sequential algorithm that lends itself to parallelization (see [22, 33, 37, 12]) finds the labels for each edge as follows. First, a spanning tree is found for the graph; the tree is then arbitrarily rooted and each vertex is assigned a level and parent. Each non-tree edge corresponds to a distinct ear, since an arbitrary spanning tree and the non-tree edges form a cycle basis for the input graph. Each non-tree edge is then examined and uniquely labeled using the level of the least common ancestor (LCA) of the edge's endpoints and a unique serial number for that edge. Finally, the tree edges are assigned ear labels by choosing the smallest label of any non-tree edge whose cycle contains it. This algorithm runs in $O((m + n) \log n)$ time.

4.2   The PRAM Algorithm

The PRAM algorithm for ear decomposition [,] is based on the second sequential algorithm. The first step computes a spanning tree in $O(\log n)$ time, using $O(n + m)$ processors. The tree can then be rooted and levels and parents assigned to nodes by using the Euler-tour technique. Labelling the nontree edges uses an LCA algorithm, which runs within the same asymptotic bounds as the first step. Next, the labels of the tree edges can be found as follows. Denote the graph by G, the spanning tree by T, the parent of a vertex v by $p(v)$, and the label of an edge (u, v) by label(u, v). For each vertex v, set $f(v) = \min\{$label$(v, u) \mid (v, u) \in G - T\}$ and $g = (x, y) \in T$, where we have $y = p(x)$. label(g) is the minimum value in the subtree rooted at x. These two substeps can be executed in $O(\log n)$ time. Finally, labelling the ears involves sorting the edges by their labels, which can be done in $O(\log n)$ time using $O(n + m)$ processors. Therefore the total running time of this CREW PRAM algorithm is $O(\log n)$ using $O(n + m)$ processors. This is not an optimal PRAM algorithm (in the sense of [ ]), because the work (processor-time product) is asymptotically greater than the sequential complexity; however, no known optimal parallel approaches are known.

In the spanning tree part of the algorithm, the vertices of the input graph G (held in shared memory) are initially assigned evenly among the processors—although the entire input graph G is of course available to every processor through the shared memory. Let D be the function on the vertex set V defining a pseudoforest (a collection of trees). Initially each vertex is in its own tree, so we set $D(v) = v$ for each $v \in V$. The algorithm manipulates the pseudoforest through two operations.

- **Grafting:** Let T_i and T_j be two distinct trees in the pseudoforest defined by D. Given the root r_i of T_i and a vertex v of T_j, the operation $D(r_i) \leftarrow v$ is called *grafting* T_i onto T_j.
- **Pointer jumping:** Given a vertex v in a tree T, the *pointer-jumping* operation applied to v sets $D(v) \leftarrow D(D(v))$.

Initially each vertex is in a rooted star. The first step will be several grafting operations of the same tree. The next step attempts to graft the rooted stars onto other trees if possible. If all vertices then reside in rooted stars, the algorithm stops. Otherwise pointer jumping is applied at every vertex, reducing the diameter of each tree, and the algorithm loops back to the first step. Figure 6 in the appendix shows the grafting and pointer-jumping operations performed on an example input graph. The ear decomposition algorithm first computes the spanning tree, then labels non-tree edges using independent LCA searches, and finally labels tree edges in parallel.

4.3   SMP Implementation and Analysis

Spanning Tree: Grafting subtrees can be carried out in $O(m/p)$ time, with two noncontiguous memory accesses to exchange approximately $\frac{n}{p}$ elements; grafting rooted stars onto other trees takes $O(m/p)$ time; pointer-jumping on all vertices takes $O(n/p)$ time; and the three steps are repeated $O(\log n)$ times. Note that the memory is accessed only once even though there are up to $\log n$ iterations. Hence the total running time of the algorithm is

$$T(n,p) = O\Big(1, \tfrac{n}{p}, ((m+n)/p)\log n\Big). \tag{1}$$

Ear Decomposition: Equation (1) gives us the running time for spanning tree formation. Computing the Euler tour takes time linear in the number of vertices per processor or $O(n/p)$, with $\frac{n}{p}$ noncontiguous memory accesses to exchange $\frac{n}{p}$ elements. The level and parent of the local vertices can be found in $O(n/p)$ time with $\frac{n}{p}$ noncontiguous memory accesses. Edge labels (for tree and nontree edges) can be found in $O(n/p)$ time. The two labeling steps need $\frac{n}{p}$ noncontiguous memory accesses to exchange approximately $\frac{n}{p}$ elements. The total running time of the algorithm is $O\Big(\frac{n}{p}, \frac{n}{p}, \frac{m+n}{p}\log n\Big)$. Since M_A and M_E are of the same order and since noncontiguous memory accesses are more expensive (due to cache faults), we can rewrite our running time as

$$T(n,p) = O\Big(\tfrac{n}{p}, \tfrac{m+n}{p}\log n\Big). \tag{2}$$

This expression *decreases* with increasing p; moreover, the algorithm scales down efficiently as well, since we have $T(n,p) = \frac{T^*(n)}{p/\log n}$, where $T^*(n)$ is the sequential complexity of ear decomposition.

5   Experimental Results

We study the performance of our implementations using a variety of input graphs that represent classes typically seen in high-performance computing applications. Our goals are to confirm that the empirical performance of our algorithms is realistically modeled and to learn what makes a parallel shared-memory implementation efficient.

5.1   Test Data

We generate graphs from seven input classes of planar graphs (2 regular and 5 irregular) that represent a diverse collection of realistic inputs. The first two classes are regular meshes (lattice RL and triangular RT); the next four classes are planar graphs generated through a simple random process, two very sparse (GA and GB) and two rather more dense (GC and GD)—since the graphs are planar, they cannot be dense in the usual sense of the word, but GD graphs are generally fully triangulated. The last graph class generates the constrained Delaunay triangulation (CD) on a set of random points [ ]. For the random graphs GA, GB, GC, and GD, the input graph on $n = |V|$ vertices is generated as follows. Random coordinates are picked in the unit square according to a uniform distribution; a Euclidean minimum spanning tree (MST) on the n vertices is formed to ensure that the graph is connected and serves as the initial edge set of the graph. Then for *count* times, two vertices are selected at random and a straight edge is added between them if the new edge does not intersect with any existing edge. Note that a count of zero produces a tree, but that the expected number of edges is generally much less than the count used in the construction, since any crossing edges will be discarded. Table 1 summarizes the seven graph families. Figure 7 in the appendix shows some example graphs with various val-

Table 1. Different classes of input graphs

Key	Name	Description
RL	Regular Lattice	Regular 4-connected mesh of $\lfloor \sqrt{n} \rfloor \times \lceil \sqrt{n} \rceil$ vertices
RT	Regular Triangulation	RL graph with an additional edge connecting a vertex to its down-and-right neighbor, if any
GA	Random Graph A	Planar, very sparse graph with $count = n$
GB	Random Graph B	Planar, very sparse graph with $count = 2n$
GC	Random Graph C	Planar graph with $count = n^2/2$
GD	Random Graph D	Planar graph with $count = n^2$
CD	Constrained Delaunay Triangulation	Constrained Delaunay triangulation of n random points in the unit square

ues of *count*. Note that, while we generate the test input graphs geometrically, only the vertex list and edge set of each graph are used in our experiments.

5.2   Test Platforms

We tested our shared-memory implementation on the NPACI Sun E10K at the San Diego Supercomputing Center. This machine has 64 processors and 64GB of shared memory, with 16KB of on-chip direct-mapped data cache and 8MB of external cache for each processor [ ].

Our practical programming environment for SMPs is based upon the SMP Node Library component of SIMPLE [ ], which provides a portable framework for describing SMP algorithms using the single-program multiple-data (SPMD) program style. This framework is a software layer built from POSIX threads that allows the user to use either already developed SMP primitives or direct thread primitives. We have been continually developing and improving this library over the past several years and have found it to be portable and efficient on a variety of operating systems (e.g., Sun Solaris, Compaq/Digital UNIX, IBM AIX, SGI IRIX, HP-UX, Linux). The SMP Node Library contains a number of SMP node algorithms for barrier synchronization, broadcasting the location of a shared buffer, replication of a data buffer, reduction, and memory management for shared-buffer allocation and release. In addition to these functions, we have control mechanisms for contextualization (executing a statement on only a subset of processors), and a *parallel do* that schedules n independent work statements implicitly to p processors as evenly as possible.

5.3   Experimental Results

Due to space limitations, we present only a few graphs illustrating our results and omit a discussion of our timing methodology. (We used elapsed times on a dedicated machine; we cross-checked our timings and ran sufficient tests to verify that our measurements did not suffer from any significant variance.) Figure 2 shows the execution time of the SMP algorithms and demonstrates that the practical performance of the SMP approach is nearly invariant with respect to the input graph class.

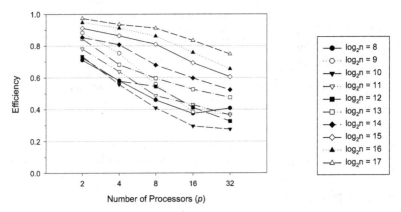

Fig. 3. Efficiency of ear decomposition for fixed inputs, each a sparse random graph with from 256 to 128K vertices, on the NPACI Sun E10K with 2 to 32 processors

The analysis for the shared-memory algorithm given in Equation (2) shows that a practical parallel algorithm is possible. We experimented with the SMP implementation on problems ranging in size from 256 to 128K vertices on the Sun E10K using $p = 1$ to 32 processors. Clearly, the nearly linear speedup with the number of processors predicted by Equation (2) may not be achievable due to synchronization overhead, serial work, or contention for shared resources. In fact, our experimental results, plotted in Figures 2, confirm the cost analysis and provide strong evidence that our shared-memory algorithm achieves nearly linear speedup with the number of processors for each fixed problem size. Figures 3 and 4 present the efficiency of our implementation when compared with the sequential algorithm for ear decomposition. In Figure 3 for a fixed problem size, as

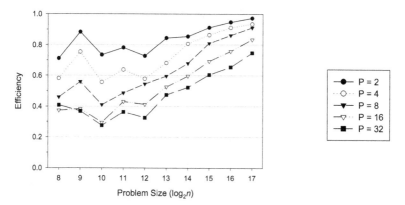

Fig. 4. Efficiency of ear decomposition for fixed machine sizes, from 2 to 32 processors, on the NPACI Sun E10K for a sparse random graph with 256 to 128K vertices

expected, the efficiency decreases as we add more processors—caused by increasing parallel overheads. Another viewpoint is that in Figure 4 for a fixed number of processors, efficiency increases with the problem size. Note that the efficiency results confirm that our implementation scales nearly linearly with the number of processors and that, as expected, larger problem sizes show better scalability.

6 Conclusions

We have implemented and analyzed a PRAM algorithm for the ear decomposition problem. We have shown both theoretically and practically that our shared-memory approach to parallel computation is efficient and scalable on a variety of input classes and problem sizes. In particular, we have demonstrated the first ever near-linear speedup for a nontrivial graph problem using a large range of processors on a shared-memory architecture. As our example shows, PRAM algorithms that once were of mostly theoretical interest, now provide plausible approaches for real implementations on UMA shared-memory architectures such as the Sun E10K. Our future work includes determining what algorithms can be efficiently parallelized in this manner on these architectures, building a basic library of efficient implementations, and using them to tackle difficult optimization problems in computational biology.

References

1. A. Aggarwal, B. Alpern, A. Chandra, and M. Snir. A Model for Hierarchical Memory. In *Proceedings of the 19th Annual ACM Symposium of Theory of Computing (STOC)*, pages 305–314, New York City, May 1987. 133

2. A. Aggarwal and J. Vitter. The Input/Output Complexity of Sorting and Related Problems. *Communications of the ACM*, 31:1116–1127, 1988. 133

3. B. Alpern, L. Carter, E. Feig, and T. Selker. The Uniform Memory Hierarchy Model of Computation. *Algorithmica*, 12:72–109, 1994. 133

4. N. M. Amato, J. Perdue, A. Pietracaprina, G. Pucci, and M. Mathis. Predicting performance on SMPs. a case study: The SGI Power Challenge. In *Proceedings of the International Parallel and Distributed Processing Symposium (IPDPS 2000)*, pages 729–737, Cancun, Mexico, May 2000. 133

5. D. A. Bader and J. JáJá. SIMPLE: A Methodology for Programming High Performance Algorithms on Clusters of Symmetric Multiprocessors (SMPs). *Journal of Parallel and Distributed Computing*, 58(1):92–108, 1999. 138

6. G. E. Blelloch, P. B. Gibbons, Y. Matias, and M. Zagha. Accounting for Memory Bank Contention and Delay in High-Bandwidth Multiprocessors. *IEEE Transactions on Parallel and Distributed Systems*, 8(9):943–958, 1997. 133

7. M. H. Carvalho, C. L. Lucchesi, and U. S. R. Murty. Ear Decompositions of Matching Covered Graphs. *Combinatorica*, 19(2):151–174, 1999. 134

8. A. Charlesworth. Starfire: extending the SMP envelope. *IEEE Micro*, 18(1):39–49, 1998. 130, 137

9. J. Chen and S. P. Kanchi. Graph Ear Decompositions and Graph Embeddings. *SIAM Journal on Discrete Mathematics*, 12(2):229–242, 1999. 134

10. P. Crescenzi, C. Demetrescu, I. Finocchi, and R. Petreschi. LEONARDO: A Software Visualization System. In *Proceedings of the First Workshop on Algorithm Engineering (WAE'97)*, pages 146–155, Venice, Italy, sep 1997. 134

11. D. Eppstein. Parallel Recognition of Series Parallel Graphs. *Information & Computation*, 98:41–55, 1992. 134

12. D. S. Franzblau. Combinatorial Algorithm for a Lower Bound on Frame Rigidity. *SIAM Journal on Discrete Mathematics*, 8(3):388–400, 1995. 134

13. D. S. Franzblau. Ear Decomposition with Bounds on Ear Length. *Information Processing Letters*, 70(5):245–249, 1999. 134

14. D. S. Franzblau. Generic Rigidity of Molecular Graphs Via Ear Decomposition. *Discrete Applied Mathematics*, 101(1-3):131–155, 2000. 134

15. P. B. Gibbons, Y. Matias, and V. Ramachandran. Can shared-memory model serve as a bridging model for parallel computation? In *Proceedings of the 9th annual ACM symposium on parallel algorithms and architectures*, pages 72–83, Newport, RI, June 1997. 133

16. P. B. Gibbons, Y. Matias, and V. Ramachandran. The Queue-Read Queue-Write PRAM Model: Accounting for Contention in Parallel Algorithms. *SIAM Journal on Computing*, 28(2):733–769, 1998. 133

17. B. Grayson, M. Dahlin, and V. Ramachandran. Experimental evaluation of QSM, a simple shared-memory model. In *Proceedings of the 13th International Parallel Processing Symposium and 10th Symposium on Parallel and Distributed Processing (IPPS/SPDP)*, pages 1–7, San Juan, Puerto Rico, April 1999. 131

18. D. R. Helman and J. JáJá. Designing Practical Efficient Algorithms for Symmetric Multiprocessors. In *Algorithm Engineering and Experimentation (ALENEX'99)*, pages 37–56, Baltimore, MD, January 1999. 133

19. T.-S. Hsu and V. Ramachandran. Efficient massively parallel implementation of some combinatorial algorithms. *Theoretical Computer Science*, 162(2):297–322, 1996. 131

20. T.-S. Hsu, V. Ramachandran, and N. Dean. Implementation of parallel graph algorithms on a massively parallel SIMD computer with virtual processing. In *Proceedings of the 9th International Parallel Processing Symposium*, pages 106–112, Santa Barbara, CA, April 1995. 131

21. L. Ibarra and D. Richards. Efficient Parallel Graph Algorithms Based on Open Ear Decomposition. *Parallel Computing*, 19(8):873–886, 1993. 134

22. J. JáJá. *An Introduction to Parallel Algorithms*. Addison-Wesley Publishing Company, New York, 1992. 130, 134, 135

23. A. Kanevsky and V. Ramachandran. Improved Algorithms for Graph Four-Connectivity. *Journal of Computer and System Sciences*, 42(3):288–306, 1991. 134

24. D. J. Kavvadias, G. E. Pantziou, P. G. Spirakis, and C. D. Zaroliagis. Hammock-On-Ears Decomposition: A Technique for the Efficient Parallel Solution of Shortest Paths and Other Problems. *Theoretical Computer Science*, 168(1):121–154, 1996. 134

25. A. Kazmierczak and S. Radhakrishnan. An Optimal Distributed Ear Decomposition Algorithm with Applications to Biconnectivity and Outerplanarity Testing. *IEEE Transactions on Parallel and Distributed Systems*, 11(1):110–118, 2000. 134

26. J. Keller, C. W. Keßler, and J. L. Träff. *Practical PRAM Programming*. John Wiley & Sons, 2001. 131

27. R. Ladner, J. D. Fix, and A. LaMarca. The cache performance of traversals and random accesses. In *Proc. 10th Ann. ACM/SIAM Symposium on Discrete Algorithms (SODA-99)*, pages 613–622, Baltimore, MD, 1999. 132

28. A. LaMarca and R. E. Ladner. The Influence of Caches on the Performance of Heaps. *ACM Journal of Experimental Algorithmics*, 1(4), 1996. http://www.jea.acm.org/1996/LaMarcaInfluence/. 132

29. A. LaMarca and R. E. Ladner. The Influence of Caches on the Performance of Heaps. In *Proceedings of the Eighth ACM/SIAM Symposium on Discrete Algorithms*, pages 370–379, New Orleans, LA, 1997. 132

30. L. Lovász. Computing Ears and Branchings in Parallel. In *Proceedings of the 26th Annual IEEE Symposium on Foundations of Computer Science (FOCS 85)*, pages 464–467, Portland, Oregon, October 1985. 134

31. Message Passing Interface Forum. MPI: A Message-Passing Interface Standard. Technical report, University of Tennessee, Knoxville, TN, June 1995. Version 1.1. 130

32. G. L. Miller and V. Ramachandran. Efficient parallel ear decomposition with applications. Manuscript, UC Berkeley, MSRI, January 1986. 135

33. Y. Moan, B. Schieber, and U. Vishkin. Parallel ear decomposition search (EDS) and st-numbering in graphs. *Theoretical Computer Science*, 47(3):277–296, 1986. 134, 135

34. B. M. E. Moret. Towards a discipline of experimental algorithmics. In *DIMACS Monographs in Discrete Mathematics and Theoretical Computer Science*. American Mathematical Society, 2001. To appear. Available at `www.cs.unm.edu/~moret/dimacs.ps`. 132

35. OpenMP Architecture Review Board. OpenMP: A Proposed Industry Standard API for Shared Memory Programming. `http://www.openmp.org/`, October 1997. 130

36. Portable Applications Standards Committee of the IEEE. *Information technology – Portable Operating System Interface (POSIX) – Part 1: System Application Program Interface (API)*, 1996-07-12 edition, 1996. ISO/IEC 9945-1, ANSI/IEEE Std. 1003.1. 130

37. V. Ramachandran. Parallel Open Ear Decomposition with Applications to Graph Biconnectivity and Triconnectivity. In J. H. Reif, editor, *Synthesis of Parallel Algorithms*, pages 275–340. Morgan Kaufman, San Mateo, CA, 1993. 134

38. V. Ramachandran. A General-Purpose Shared-Memory Model for Parallel Computation. In M. T. Heath, A. Ranade, and R. S. Schreiber, editors, *Algorithms for Parallel Processing*, volume 105, pages 1–18. Springer-Verlag, New York, 1999. 133

39. M. Reid-Miller. List ranking and list scan on the Cray C-90. In *Proceedings Symposium on Parallel Algorithms and Architectures*, pages 104–113, Cape May, NJ, June 1994. 131

40. M. Reid-Miller. List ranking and list scan on the Cray C-90. *Journal of Computer and System Sciences*, 53(3):344–356, December 1996. 131

41. J. H. Reif, editor. *Synthesis of Parallel Algorithms*. Morgan Kaufmann Publishers, 1993. 130

42. C. Savage and J. JáJá. Fast, Efficient Parallel Algorithms for Some Graph Problems. *SIAM Journal on Computing*, 10(4):682–691, 1981. 134

43. J. R. Shewchuk. Triangle: Engineering a 2D Quality Mesh Generator and Delaunay Triangulator. In M. C. Lin and D. Manocha, editors, *Applied Computational Geometry: Towards Geometric Engineering*, volume 1148 of *Lecture Notes in Computer Science*, pages 203–222. Springer-Verlag, May 1996. From the First ACM Workshop on Applied Computational Geometry. 137

44. J. Sibeyn. Better trade-offs for parallel list ranking. In *Proceedings of the 9th annual ACM symposium on parallel algorithms and architectures*, pages 221–230, Newport, RI, June 1997. 131

45. J. Vitter and E. Shriver. Algorithms for Parallel Memory I: Two-Level Memories. *Algorithmica*, 12:110–147, 1994. 133

46. H. Whitney. Non-Separable and Planar Graphs. *Transactions of the American Mathematical Society*, 34:339–362, 1932. 134

Appendix: Illustrations

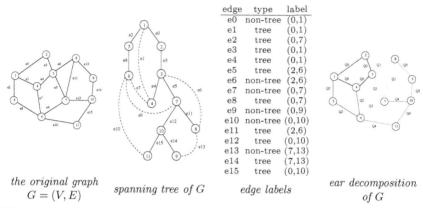

edge	type	label
e0	non-tree	(0,1)
e1	tree	(0,1)
e2	tree	(0,7)
e3	tree	(0,1)
e4	tree	(0,1)
e5	tree	(2,6)
e6	non-tree	(2,6)
e7	non-tree	(0,7)
e8	tree	(0,7)
e9	non-tree	(0,9)
e10	non-tree	(0,10)
e11	tree	(2,6)
e12	tree	(0,10)
e13	non-tree	(7,13)
e14	tree	(7,13)
e15	tree	(0,10)

the original graph
$G = (V, E)$ *spanning tree of G*   *edge labels*   *ear decomposition of G*

Fig. 5. A graph, a spanning tree for that graph, edge classification and labels, and an ear decomposition for that graph

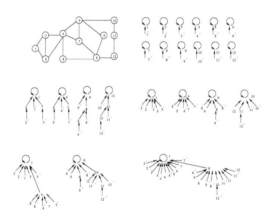

Fig. 6. Grafting and pointer-jumping operations applied to a sample graph (top left) and steps of the algorithm

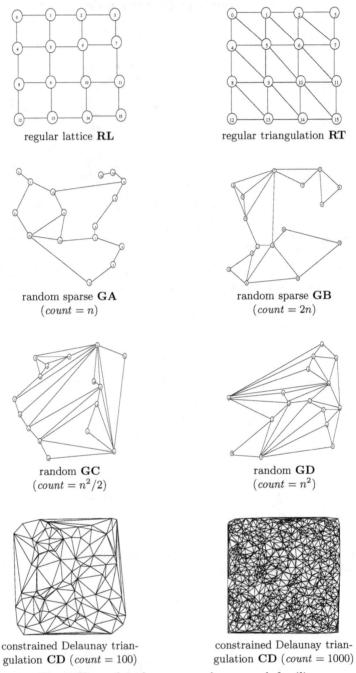

regular lattice **RL**

regular triangulation **RT**

random sparse **GA**
($count = n$)

random sparse **GB**
($count = 2n$)

random **GC**
($count = n^2/2$)

random **GD**
($count = n^2$)

constrained Delaunay triangulation **CD** ($count = 100$)

constrained Delaunay triangulation **CD** ($count = 1000$)

Fig. 7. Examples of our seven planar graph families

An Experimental Study of Data Migration Algorithms

Eric Anderson[1], Joe Hall[2], Jason Hartline[2], Michael Hobbs[1], Anna R. Karlin[2], Jared Saia[2], Ram Swaminathan[1], and John Wilkes[1]

[1] Storage Systems Program, Hewlett-Packard Laboratories
Palo Alto, CA 94304
{anderse,mjhobbs,swaram,wilkes}@hpl.hp.com
[2] Department of Computer Science and Engineering, University of Washington
Seattle, WA 98195
{jkh,hartline,karlin,saia}@cs.washington.edu

Abstract. The *data migration* problem is the problem of computing a plan for moving data objects stored on devices in a network from one configuration to another. Load balancing or changing usage patterns might necessitate such a rearrangement of data. In this paper, we consider the case where the objects are fixed-size and the network is complete. We introduce two new data migration algorithms, one of which has provably good bounds. We empirically compare the performance of these new algorithms against similar algorithms from Hall et al. [ ] which have better theoretical guarantees and find that in almost all cases, the new algorithms perform better. We also find that both the new algorithms and the ones from Hall et al. perform much better in practice than the theoretical bounds suggest.

1 Introduction

The performance of modern day large-scale storage systems (such as disk farms) can be improved by balancing the load across devices. Unfortunately, the optimal data layout is likely to change over time because of workload changes, device additions, or device failures. Thus, it may be desirable to periodically compute a new assignment of data to devices [, , ,], either at regular intervals or on demand as system changes occur. Once the new assignment is computed, the data must be migrated from the old configuration to the new configuration. This migration must be done as efficiently as possible to minimize the impact of the migration on the system. The large size of the data objects (gigabytes are common) and the the large amount of total data (can be petabytes) makes migration a process which can easily take several days.

In this paper, we consider the problem of finding an efficient migration plan. We focus solely on the offline migration problem i.e. we ignore the load imposed by user requests for data objects during the migration. Our motivation for studying this problem lies in migrating data for large-scale storage system management tools such as *Hippodrome* [ ]. Hippodrome automatically adapts to

G. Brodal et al. (Eds.): WAE 2001, LNCS 2141, pp. 145–158, 2001.

changing demands on a storage system without human intervention. it analyzes a running workload of requests to data objects, calculates a new load-balancing configuration of the objects and then migrates the objects. An offline migration can be performed as a background process or at a time when loads from user requests are low (e.g. over the weekend).

The input to the *migration problem* is an initial and final configuration of data objects on devices, and a description of the storage system (the storage capacity of each device, the underlying network connecting the devices and the size of each object.) Our goal is to find a *migration plan* that uses the existing connections between storage devices to move the data from the initial configuration to the final configuration in as few time steps as possible. We assume that the objects are all the same size, for example, fragmenting them into fixed sized extents. We also assume that any pair of devices can send to each other without impacting other pairs, i.e., we assume that the underlying network is complete. A crucial constraint on the legal parallelism in any plan is that each storage device can be involved in the transfer (either sending or receiving, but not both) of only one object at a time.

We consider two interesting variants of the fixed size migration problem. First, we consider the effect of space constraints. For *migration without space constraints* we assume that an unlimited amount of storage for data objects is available at each device in the network. At the other extreme, in *migration with space constraints*, we assume that each device has the minimum amount of extra space possible – only enough to hold one more object than the maximum of the number of objects stored on that device in the initial configuration or in the final configuration. At the end of each step of a migration plan, we ensure that the number of objects stored at a node is no more than the assumed total space at that node.

Second, we compare algorithms that migrate data directly from source to destination with those that allow indirect migration of data, through intermediate nodes. We call these intermediate nodes *bypass nodes*. Frequently, there are devices with no objects to send or receive, and we can often come up with a significantly faster migration plan if we use these devices as bypass nodes.

We can model the input to our problem as a directed *multigraph*[1] $G = (V, E)$ without self-loops that we call the *demand graph*. Each of the vertices in the demand graph corresponds to a storage device, and each of the directed edges (u, v) represents an object that must be moved from storage device u (in the initial configuration) to storage device v (in the final configuration). The output of our algorithms will be a positive integer label for each edge which indicates the stage at which that edge is moved. I/O constraints imply that no vertex can have two edges with the same integer label incident to it.

The labels on the edges may be viewed as colors in an edge coloring of the graph. Thus, direct migration with no space constraints is equivalent to the well known multigraph edge-coloring problem. The minimum number of colors needed to edge-color a graph is called the *chromatic index* or χ' of the graph. Computing

---

[1] A multigraph is a graph which can have multiple edges between any two nodes.

χ' is NP-complete but there is a $1.1\chi'(G) + .8$ approximation algorithm [ ]. Δ, the maximum degree of the graph, is a trivial lower bound on the number of colors needed. It is also well known that 1.5Δ colors always suffice and that there are graphs requiring this many colors.

For indirect migration, we want to get as close as possible to the theoretically minimum-length migration plan of Δ while minimizing the number of bypass nodes needed. The following example shows how a bypass node can be used to reduce the number of stages. In the graph on the left, each edge is duplicated k times and clearly $\chi' = 3k$. However, using only one bypass node, we can perform the migration in $\Delta = 2k$ stages as shown on the right. (The bypass node is shown as o.)

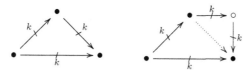

It is easy to see that $n/3$ is a *worst case lower bound* on the number of bypass nodes needed to perform a migration in Δ stages – consider the example demand graph consisting of k disjoint copies of the 3-cycle ($n = 3k$).

In this paper, we introduce two new migration algorithms. The primary focus of our work is on the empirical evaluation of these algorithms, and the migration algorithms introduced in [ ].

For the case where there are no space constraints, we evaluate two algorithms which use indirection. We introduce the first of these in this paper; it is called *Max-Degree-Matching*. This algorithm can find a migration taking Δ steps while using no more than $2n/3$ bypass nodes. We compare this to *2-factoring* [ ] which finds a migration plan taking $2\lceil \Delta/2 \rceil$ steps by using no more than $n/3$ bypass nodes [ ]. While *2-factoring* has better theoretical bounds than *Max-Degree-Matching*, we will see that *Max-Degree-Matching* uses fewer bypass nodes on almost all tested demand graphs.

For migration with space constraints, we introduce a new algorithm, *Greedy-Matching*, which uses no bypass nodes. We know of no good bound on the number of time steps taken by *Greedy-Matching* in the worst case; however, in our experiments, *Greedy-Matching* often returned plans with very close to Δ time steps and never took more than $3\Delta/2$ time steps. This compares favorably with *4-factoring direct* [ ] which also never uses bypass nodes but which always takes essentially $3\Delta/2$ time steps.

The paper is organized as follows. In Section 2, we describe the algorithms we have evaluated for indirect migration without space constraints. In Section 3, we describe the algorithms we have evaluated for migration with space constraints. Section 4 describes how we create the demand graphs on which we test the migration algorithms while Sections 5 and 6 describe our experimental results. Section 7 gives an analysis and discussion of these results and Section 8 summarizes and gives directions for future research.

2   Indirect Migration without Space Constraints

We begin with a new algorithm, *Max-Degree-Matching* which uses at most $2n/3$ bypass nodes and always attains an optimal Δ step migration plan without space constraints. *Max-Degree-Matching* works by sending, in each stage, one object from each vertex in the demand graph that has maximum degree. To do this, we first find a matching which matches all maximum-degree vertices with no out-edges. Next, we match each unmatched maximum-degree vertex up with a bypass node. Finally we use the general matching algorithm [ ] to expand this matching to a maximal matching and then send every edge in this new expanded matching. The full algorithm is given in Appendix A.1; a proof of the following theorem is given in [ ].

Theorem 1. Max-Degree-Matching *computes a correct Δ-stage migration plan using at most $2n/3$ bypass nodes.*

We compare *Max-Degree-Matching* with *2-factoring* from Hall et al. which also computes an indirect migration plan without space constraints. Hall et al., show that *2-factoring* takes $2\lceil\Delta/2\rceil$ time steps while using no more than $n/3$ bypass nodes.

We note that in a particular stage of *2-factoring* as described in Hall et al., there may be some nodes which only have dummy edges incident to them. A heuristic for reducing the number of bypass nodes needed is to use these nodes as bypass nodes when available to decrease the need for "external" bypass nodes. Our implementation of *2-factoring* uses this heuristic.

3   Migration with Space Constraints

The *Greedy Matching* algorithm (Algorithm 1) is a new and straightforward direct migration algorithm which obeys space constraints. This algorithm eventually sends all of the objects [ ] but the worst case number of stages is unknown.

Algorithm 1 *Greedy Matching*

---

1. Let G' be the graph induced by the sendable edges in the demand graph. An edge is sendable if there is free space at its destination.
2. Compute a maximum general matching on G'.
3. Schedule all edges in matching to be sent in this stage.
4. Remove these edges from the demand graph.
5. Repeat until the demand graph has no more edges.

We compare *Greedy-Matching* with two provably good algorithms for migration with space constraints from Hall et al.. We refer to these algorithms as

4-*factoring direct* and 4-*factoring indirect*. Hall et al. show that 4-*factoring direct* finds a $6\lceil\Delta/4\rceil$ stage migration without bypass nodes and that *4-factoring indirect* finds a $4\lceil\Delta/4\rceil$ stage migration plan using at most $n/3$ bypass nodes.

In our implementation of 4-*factoring indirect*, we again use the heuristic of using nodes with only dummy edges in a particular stage as bypass nodes for that stage.

4   Experimental Setup

The following table summarizes the theoretical results known for each algorithm[2].

Algorithm	Type	Space Constraints	Plan Length	Worst Case Max. Bypass Nodes
2-factoring [ ]	indirect	No	$2\lceil\Delta/2\rceil$	$n/3$
Max-Degree-Matching	indirect	No	Δ	$2n/3$
Greedy-Matching	direct	Yes	unknown	0
4-factoring direct [ ]	direct	Yes	$6\lceil\Delta/4\rceil$	0
4-factoring indirect [ ]	indirect	Yes	$4\lceil\Delta/4\rceil$	$n/3$

We tested these algorithms on four types of multigraphs[3]:

1. *Load-Balancing Graphs.* These graphs represent real-world migrations. A detailed description of how they were created is given in the next subsection.
2. *General Graphs*(n, m). A graph in this class contains n nodes and m edges. The edges are chosen uniformly at random from among all possible edges disallowing self-loops (but allowing parallel edges).
3. *Regular Graphs*(n, d). Graphs in this class are chosen uniformly at random from among all regular graphs with n nodes, where each node has total degree d (where d is even). We generated these graphs by taking the edge-union of $d/2$ randomly generated 2-regular graphs over n vertices.
4. *Zipf Graphs*(n, d_{min}). These graphs are chosen uniformly at random from all graphs with n nodes and minimum degree d_{min} that have Zipf degree distribution i.e. the number of nodes of degree d is proportional to $1/d$. Our technique for creating random Zipf graphs is given in detail in [ ].

4.1   Creation of Load-Balancing Graphs

A migration problem can be generated from any pair of configurations of objects on nodes in a network. To generate the *Load-Balancing* graphs, we used two different methods of generating sequences of configurations of objects which might

---

[2] For each algorithm, time to find a migration plan is negligible compared to time to implement the plan.

[3] Java code implementing these algorithms along with input files for all the graphs tested is available at www.cs.washington.edu/homes/saia/migration

occur in a real world system. For each sequence of say l configurations, $C_1, \ldots C_l$, for each i, $1 \leq i \leq l-1$, we generate a demand graph using C_i as the initial configuration and C_{i+1} as the final.

For the first method, we used the Hippodrome system on two variants of a retail data warehousing workload [ ]. Hippodrome adapts a storage system to support a given workload by generating a series of object configurations, and possibly increasing the node count. Each configuration is better at balancing the workload of user queries for data objects across the nodes in the network than the previous configuration. We ran the Hippodrome loop for 7 iterations (enough to stabilize the node count) and so got two sequences of 7 configurations. For the second method, we took the 17 queries to a relational database in the TPC-D benchmark [ ] and for each query generated a configuration of objects to devices which balanced the load across the devices effectively. This method gives us a sequence of 17 configurations.

Different devices have different performance properties and hence induce different configurations. When generating our configurations, we assumed all nodes in the network were the same device. For both methods, we generated configurations based on two different types of devices. Thus for the Hippodrome method, we generated 4 sequences of 7 configurations (6 graphs) and for the TPC-D method, we generated 2 sequences of 17 configurations (16 graphs) for a total of 56 demand graphs.

5   Results on the *Load-Balancing Graphs*

5.1   Graph Characteristics

Detailed plots on the characteristics of the *load-balancing* graphs are given in [ ] and are summarized here briefly. We refer to the sets of graphs generated by Hippodrome on the first and second device type as the first and second set respectively and the sets of graphs generated with the TPC-D method for the first and second device types as the third and fourth sets.

The number of nodes in each graph is less than 50 for the graphs in all sets except the third in which most graphs have around 300 nodes. The edge density for each graph varies from about 5 for most graphs in the third set to around 65 for most graphs in the fourth set. The Δ value for each graph varies from about 15 to about 140, with almost all graphs in the fourth set having density around 140.

5.2   Performance

Figure 1 shows the performance of the algorithms on the load-balancing graphs in terms of the number of bypass nodes used and the time steps taken. The x-axis in each plot gives the index of the graph which is consistent across both plots. The indices of the graphs are clustered according to the sets the graphs are from with the first, second, third and fourth sets appearing left to right, separated by solid lines.

The first plot shows the number of bypass nodes used by *2-factoring*, *4-factoring indirect* and *Max-Degree-Matching*. We see that *Max-Degree-Matching* uses 0 bypass nodes on most of the graphs and never uses more than 1. The number of bypass nodes used by *2-factoring* and *4-factoring indirect* are always between 0 and 6, even for the graphs with about 300 nodes. The second plot shows the number of stages required divided by Δ for *Greedy-Matching*. Recall that this ratio for *2-factoring,Max-Degree-Matching* and *4-factoring indirect* is essentially 1 while the ratio for *4-factoring direct* is essentially 1.5. In the graphs in the second and third set, *Greedy-Matching* almost always has a ratio near 1. However in the first set, *Greedy-Matching* has a ratio exceeding 1.2 on several of the graphs and a ratio of more than 1.4 on one of them. In all cases, *Greedy-Matching* has a ratio less than *4-factoring direct*.

We note the following important points: (1) On all of the graphs, the number of bypass nodes needed is less than 6 while the theoretical upper bounds are significantly higher. In fact, *Max-Degree-Matching* used *no bypass nodes* for the majority of the graphs (2) *Greedy-Matching* always takes fewer stages than *4-factoring direct*.

6   Results on General, Regular and Zipf Graphs

6.1   Bypass Nodes Needed

For *General, Regular* and *Zipf Graphs*, for each set of graph parameters tested, we generated 30 random graphs and took the average performance of each algorithm over all 30 graphs. For this reason, the data points in the plots are not at integral values. *Greedy-Matching* never uses any bypass nodes so in this section, we include results only for *Max-Degree-Matching*, *4-factoring indirect* and *2-factoring*.

Varying Number of Nodes The three plots in the left column of Figure 2 give results for random graphs where the edge density is fixed and the number of nodes varies. The first plot in this column shows the number of bypass nodes used for *General Graphs* with edge density fixed at 10 as the number of nodes increases from 100 to 1200. We see that *Max-Degree-Matching* and *2-factoring* consistently use no bypass nodes. *4-factoring indirect* uses between 2 and 3 bypass nodes and surprisingly this number does not increase as the number of nodes in the graph increases.

The second plot shows the number of bypass nodes used for *Regular Graphs* with $\Delta = 10$ as the number of nodes increases from 100 to 1200. We see that the number of bypass nodes needed by *Max-Degree-Matching* stays relatively constant at 1 as the number of nodes increases. The number of bypass nodes used by *2-factoring* and *4-factoring indirect* are very similar, starting at 3 and growing very slowly to 6, approximately linearly with a slope of 1/900.

The third plot shows the number of bypass nodes used on *Zipf Graphs* with minimum degree 1 as the number of nodes increases. In this graph, *2-factoring*

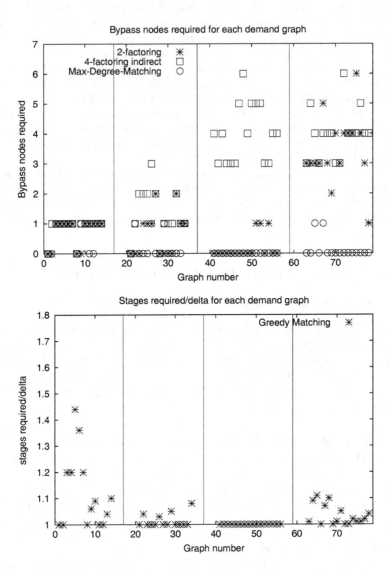

Fig. 1. The top plot gives the number of bypass nodes required for the algorithms *2-factoring*, *4-factoring indirect* and *Max-Degree-Matching* on each of the *Load-Balancing Graphs*. The bottom plot gives the ratio of time steps required to Δ for *Greedy-Matching* on each of the *Load-Balancing Graphs*. The three solid lines in both plots divide the four sets of *Load-Balancing Graphs*

is consistently at 0, *Max-Degree-Matching* varies between 1/4 and 1/2 and 4-*factoring indirect* varies between 1 and 4.

Varying Edge Density The three plots in the right column of Figure 2 show the number of bypass nodes used for graphs with a fixed number of nodes as the edge density varies. The first plot in the column shows the number of bypass nodes used on *General Graphs*, when the number of nodes is fixed at 100, and edge density is varied from 20 to 200. We see that the number of bypass nodes used by *Max-Degree-Matching* is always 0. The number of bypass nodes used by 2 and 4-*factoring indirect* increases very slowly, approximately linearly with a slope of about 1/60. Specifically, the number used by 2-factoring increases from 1/2 to 6 while the number used by 4-*factoring indirect* increases from 4 to 6.

The second plot shows the number of bypass nodes used on *Regular Graphs*, when the number of nodes is fixed at 100 and Δ is varied from 20 to 200. The number of bypass nodes used by *Max-Degree-Matching* stays relatively flat varying slightly between 1/2 and 1. The number of bypass nodes used by 2-*factoring* and 4-*factoring indirect* increases near linearly with a larger slope of 1/30, increasing from 4 to 12 for 2-*factoring* and from 4 to 10 for 4-*factoring indirect*.

The third plot shows the number of bypass nodes used on *Zipf Graphs*, when the number of nodes is fixed at 146 and the minimum degree is varied from 1 to 10. 2-*factoring* here again always uses 0 bypass nodes. The *Max-Degree-Matching* curve again stays relatively flat varying between 1/4 and 1. 4-*factoring indirect* varies slightly, from 2 to 4, again near linearly with a slope of 1/5.

We suspect that our heuristic of using nodes with only dummy edges as bypass nodes in a stage helps 2-*factoring* significantly on *Zipf Graphs* since there are so many nodes with small degree and hence many dummy self-loops.

6.2   Time Steps Needed

For *General* and *Regular Graphs*, the migration plans *Greedy-Matching* found never took more than $\Delta + 1$ time steps. Since the other algorithms we tested are guaranteed to have plans taking less than $\Delta + 3$, we present no plots of the number of time steps required for these algorithms on *General* and *Regular Graphs*.

As shown in Figure 3, the number of stages used by *Greedy-Matching* for *Zipf Graphs* is significantly worse than for the other types of random graphs. We note however that it always performs better than 4-*factoring direct*. The first plot shows that the number of extra stages used by *Greedy-Matching* for *Zipf Graphs* with minimum degree 1 varies from 2 to 4 as the number of nodes varies from 100 to 800. The second plot shows that the number of extra stages used by *Greedy-Matching* for *Zipf Graphs* with 146 nodes varies from 1 to 11 as the minimum degree of the graphs varies from 1 to 10. High density Zipf graphs are the one weakness we found for *Greedy-Matching*.

154   Eric Anderson et al.

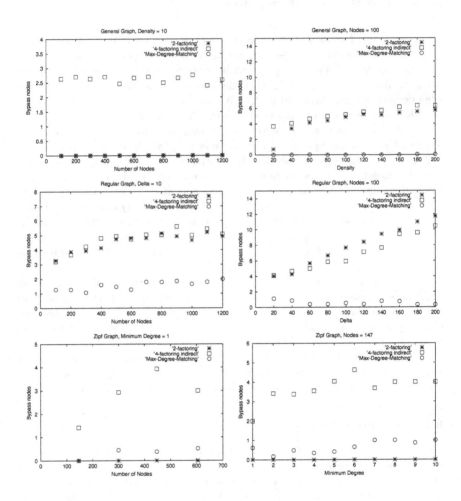

Fig. 2. These six plots give the number of bypass nodes needed for *2-factoring*, *4-factoring direct* and *Max-Degree-Matching* for the *General, Regular* and *Zipf Graphs*. The three plots in the left column give the number of bypass nodes needed as the number of *nodes* in the random graphs increase. The three plots in the right column give the number of bypass nodes needed as the *density* of the random graphs increase. The plots in the first row are for *General Graphs*, plots in the second row are for *Regular Graphs* and plots in the third row are for *Zipf Graphs*

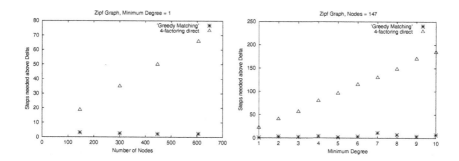

Fig. 3. Number of steps above *Delta* needed for *Greedy-Matching* on *Zipf Graphs*

7   Analysis

Our major empirical conclusions for the graphs tested are:

- *Max-Degree-Matching* almost always uses less bypass nodes than *2-factoring*.
- *Greedy-Matching* always takes less time steps than *4-factoring direct*.
- For all algorithms using indirection, the number of bypass nodes required is almost always no more than $n/30$.

For migration without space constraints, *Max-Degree-Matching* performs very well in practice, often using significantly fewer bypass nodes than *2-factoring*. Its good performance and good theoretical properties make it an attractive choice for real world migration problems without space constraints.

For migration with space constraints, *Greedy-Matching* always outperforms *4-factoring direct*. It also frequently finds migration plans within some small constant of Δ. However there are many graphs for which it takes much more than Δ time steps and for this reason we recommend *4-factoring indirect* when there are bypass nodes available.

7.1   Theory versus Practice

In our experiments, we have found that not only are the number of bypass nodes required for the types of graphs we tested much less than the theoretical bounds suggest but that in addition, the *rate* of growth in the number of bypass nodes versus the number of demand graph nodes is much less than the theoretical bounds. The worst case bounds are that $n/3$ bypass nodes are required for *2-factoring* and *4-factoring indirect* and $2n/3$ for *Max-Degree-Matching* but in most graphs, for all the algorithms, we never required more than about $n/30$ bypass nodes.

The only exception to this trend is regular graphs with high density for which *2-factoring* and *4-factoring indirect* required up to $n/10$ bypass nodes. A surprising result for these graphs was the fact that *Max-Degree-Matching* performed so much better than *2-factoring* and *4-factoring indirect* despite its worse theoretical bound.

8   Conclusion

We have introduced two new data migration algorithms and have empirically evaluated their performance compared with two algorithms from [ ]. The metrics we used to evaluate the algorithms are: (1) the number of time steps required to perform the migration, and (2) the number of bypass nodes used as intermediate storage devices. We have found on several types of random and load-balancing graphs that the new algorithms outperform the algorithms from [ ] on these two metrics despite the fact that the theoretical bounds for the new algorithms are not as good. Not surprisingly, we have also found that for all the algorithms tested, the theoretical bounds are overly pessimistic. We conclude that many of the algorithms described in this paper are both practical and effective for data migration.

There are several directions for future work. Real world devices have different I/O speeds. For example, one device might be able handle sending or receiving twice as many objects per stage as another device. We want good approximation algorithms for migration with different device speeds. Also in some important cases, a complete graph is a poor approximation to the network topology. For example, a wide area network typically has a very sparse topology which is more closely related to a tree than to a complete graph. We want good approximation algorithms for commonly occuring topologies (such as trees) and in general for arbitrary topologies. Saia [ ] gives some preliminary approximation algorithms for migration with variable device speeds and different network topologies.

Another direction for future work is designing algorithms which work well for the online migration problem. In this paper, we have ignored loads imposed by user requests in devising a migration plan. A better technique for creating a migration plan would be to migrate the objects in such a way that we interfere as little as possible with the ability of the devices to satisfy user requests and at the same time improve the load balancing behavior of the network as quickly as possible. This may require adaptive algorithms since user requests are unpredictable.

A final direction for future work is designing algorithms which make use of free nodes when available but do not *require* them to perform well. In particular, we want a good approximation algorithm for migration with indirection when no external bypass nodes are available. To the best of our knowledge, no algorithm with an approximation ratio better than 3/2 for this problem is known at this time.

References

1. E. Anderson, J. Hall, J. Hartline, M. Hobbs, A. Karlin, J. Saia, R. Swaminathan, and J. Wilkes. An Experimental Study of Data Migration Algorithms. Technical report, University of Washington, 2001. 148, 149, 150
2. E. Anderson, M. Hobbs, K. Keeton, S. Spence, M. Uysal, and A. Veitch. Hippodrome: running circles around storage administration. Submitted to Symposium on Operating System Principles, 2001. 145, 150
3. E. Borowsky, R. Golding, A. Merchant, L. Schreier, E. Shriver, M. Spasojevic, and J. Wilkes. Using attribute-managed storage to achieve QoS. In *5th Intl. Workshop on Quality of Service*, Columbia Univ., New York, June 1997. 145
4. Transaction Processing Performance Council. TPC Benchmark D (Decision Support) Standard Specification Revision 2.1. 1996. 150
5. B. Gavish and O. R. Liu Sheng. Dynamic file migration in distributed computer systems. *Communications of the ACM*, 33:177–189, 1990. 145
6. I. Golubchik, S. Khuller S. Khanna, R. Thurimella, and A. Zhu. Approximation Algorithms for Data Placement on Parallel Disks. In *Proceedings of the Eleventh Annual ACM-SIAM Symposium on Discrete ALgorithms*, pages 223–232, 2000. 145
7. J. Hall, J. Hartline, A. Karlin, J. Saia, and J. Wilkes. On algorithms for efficient data migration. In *12th annual ACM-SIAM Symposium on Discrete Algorithms*, 2001. 145, 147, 149, 156
8. S. Micali and V. Vazirani. An $O(\sqrt{|V|}|E|)$ algorithm for finding a maximum matching in general graphs. In *21st Annual Symposium on Foundations of Computer Science*, 1980. 148, 158
9. T. Nishizeki and K. Kashiwagi. On the 1.1 edge-coloring of multigraphs. In *SIAM Journal on Discrete Mathematics*, volume 3, pages 391–410, 1990. 147
10. J. Saia. Data Migration with Edge Capacities and Machine Speeds. Technical report, University of Washington, 2001. 156
11. J. Wolf. The Placement Optimization Problem: a practical solution to the disk file assignment problem. In *Proceedings of the ACM SIGMETRICS international conference on Measurement and modeling of computer systems*, pages 1–10, 1989. 145

A   Appendix

A.1   *Max-Degree-Matching*

Algorithm 2 *Max-Degree-Matching(demand graph G)*

---

1. Set up a bipartite matching problem as follows: the left hand side of the graph is all maximum degree vertices *not adjacent to degree one vertices* in G, the right hand side is all their neighbors in G, and the edges are all edges between maximum degree vertices and their neighbors in G .
2. Find the maximum bipartite matching. The solution induces cycles and paths in the demand graph. All cycles contain only maximum degree vertices, all paths have one endpoint that is not a maximum degree vertex.
3. Mark every other edge in the cycles and paths. For odd length cycles, one vertex will be left with no marked edges. Make sure that this is a vertex with an outgoing edge (and thus can be bypassed if needed). Each vertex has at most one edge marked. Mark every edge between a maximum degree vertex and a degree one vertex.
4. Let V' be the set of vertices incident to a marked edge. Compute a maximum matching in G that matches all vertices in V' (This can be done by seeding the general matching algorithm [ ] with the matching that consists of marked edges.) Define S to be all edges in the matching.
5. For each edge vertex u of maximum degree with no incident edge in S, let (u, v) be some out-edge from u. Add (u, b) to S, where b is an unused bypass node, and add (b, v) to the demand graph G.
6. Schedule all edges in S to be sent in the next stage and remove these edges from the demand graph.
7. If there are still edges in the demand graph, go back to step 1.

An Experimental Study of Basic Communication Protocols in Ad-hoc Mobile Networks*

Ioannis Chatzigiannakis[1,2], Sotiris Nikoletseas[1,2], Nearchos Paspallis[2], Paul Spirakis[1,2], and Christos Zaroliagis[1,2]

[1] Computer Technology Institute
P.O. Box 1122, 26110 Patras, Greece
{ichatz,nikole,spirakis}@cti.gr
[2] Department of Computer Engineering and Informatics, University of Patras
26500 Patras, Greece
{paspalis,zaro}@ceid.upatras.gr

Abstract. We investigate basic communication protocols in ad-hoc mobile networks. We follow the semi-compulsory approach according to which a small part of the mobile users, the *support* Σ, that moves in a predetermined way is used as an intermediate pool for receiving and delivering messages. Under this approach, we present a new semi-compulsory protocol called the *runners* in which the members of Σ perform concurrent and continuous random walks and exchange any information given to them by senders when they meet. We also conduct a comparative experimental study of the runners protocol with another existing semi-compulsory protocol, called the *snake*, in which the members of Σ move in a coordinated way and always remain pairwise adjacent. The experimental evaluation has been carried out in a new generic framework that we developed to implement protocols for mobile computing. Our experiments showed that for both protocols only a small support is required for efficient communication, and that the runners protocol outperforms the snake protocol in almost all types of inputs we considered.

1 Introduction

Mobile computing has been introduced in the past few years as a new computing environment. Since mobile computing is constrained by poor resources, highly dynamic variable connectivity, and volatile energy sources, the design of stable and efficient mobile information systems is greatly complicated. Until now, two basic models have been proposed for mobile computing. The earlier (and commonly used) model is the *fixed backbone* model which assumes that an existing infrastructure of support stations with centralized network management is provided in order to ensure efficient communication. A more recent model is

* This work was partially supported by the IST Programme of EU under contract no. IST-1999-14186 (ALCOM-FT), by the Human Potential Programme of EU under contracts no. HPRN-CT-1999-00104 (AMORE) and HPRN-CT-1999-00112 (ARACNE), and by the Greek GSRT project ALKAD.

G. Brodal et al. (Eds.): WAE 2001, LNCS 2141, pp. 159–171, 2001.

the *ad-hoc model* which assumes that mobile hosts can form networks without the participation of any fixed infrastructure. An *ad-hoc mobile network* [,] is a collection of mobile hosts with wireless network interfaces forming a temporary network without the aid of any established infrastructure or centralized administration. In an ad-hoc network two hosts that want to communicate may not be within wireless transmission range of each other, but could communicate if other hosts between them are also participating in the ad-hoc network and are willing to forward packets for them.

A usual scenario that motivates the ad-hoc mobile model is the case of rapid deployment of mobile hosts in an unknown terrain, where there is no underlying fixed infrastructure either because it is impossible or very expensive to create such an infrastructure, or because it is not established yet, or it has become temporarily unavailable (i.e., destroyed or down).

A *basic communication problem* in such ad-hoc mobile networks, is to send information from some *sender* S, to another designated *receiver* R. Note that ad-hoc mobile networks are dynamic in nature, in the sense that local connections are temporary and may change as users move. The movement rate of each user might vary, while certain hosts might even stop in order to execute location-oriented tasks (e.g., take measurements). In such an environment, executing a distributed protocol has certain complications, since communication between two hosts may be a highly non-trivial task.

The most common way to establish communication is to form paths of intermediate nodes (i.e., hosts), where it is assumed that there is a link between two nodes if the corresponding hosts lie within one another's transmission radius and hence can directly communicate with each other [, ,]. Indeed, this approach of exploiting pairwise communications is common in ad-hoc mobile networks that either cover a relatively small space (i.e., the temporary network has a small diameter with respect to the transmission range), or are dense (i.e., thousands of wireless nodes). Since almost all locations are occupied by some hosts, broadcasting can be efficiently accomplished.

In wider area ad-hoc networks however, broadcasting is impractical, as two distant peers will not be reached by any broadcast since users do not occupy all intervening locations, i.e., a sufficiently long communication path is difficult to establish. Even if such a path is established, single link failures happening when a small number of users that were part of the communication path move in a way such that they are no longer within the transmission range of each other, will make this path invalid. Note also that the path established in this way may be very long, even in the case of connecting nearby nodes.

A different approach to solve this basic communication problem is to take advantage of the mobile hosts natural movement by exchanging information whenever mobile hosts meet incidentally. When the users of the network meet often and are spread in a geographical area, flooding the network will suffice. It is evident, however, that if the users are spread in remote areas and they do not move beyond these areas, there is no way for information to reach them, unless the protocol takes care of such situations.

One way to alleviate these problems is to force mobile users to move according to a specific scheme in order to meet the protocol demands. Our approach is based on the idea of forcing only a very small subset of mobile users, called the *support* Σ of the network, to move as per the needs of the protocol. Such protocols are called *semi-compulsory* protocols.

The first semi-compulsory protocol for the basic communication problem was presented in [ ]. It uses a snake-like sequence of support stations that always remain pairwise adjacent and move in a way determined by the snake's head. The head moves by executing a random walk over the area covered by the network. We shall refer to this protocol as the *snake protocol*. The snake protocol is theoretically analyzed in [ ] using interesting properties of random walks and their meeting times. A first implementation of the protocol was developed and experimentally evaluated in [ ] with the emphasis to confirm the theoretical analysis. Both the experiments and the theoretical analysis indicated that only a small support is needed for efficient communication.

In this paper we firstly present a new semi-compulsory protocol for the basic communication problem studied here. The new protocol is based on the idea that the members of Σ move like "runners", i.e., they move independently of each other sweeping the whole area covered by the network. When two runners meet, they exchange information given to them by senders encountered. We shall refer to this protocol as the *runners protocol*. The new protocol turns out to be more robust than the snake protocol. The latter is resilient only to one fault (one faulty member of Σ), while the former is resilient to t faults for any $0 < t < |\Sigma|$.

The second contribution of this paper is a comparative experimental study of the snake and the runners protocols based on a new generic framework that we developed to implement protocols for mobile computing. Based on the implementation in [ ], we have redesigned the fundamental classes and their functionality in order to provide this generic framework that allows implementation of any mobile protocol. Under this framework, we have implemented the runners protocol and re-implemented the snake protocol. All of our implementations were based on the LEDA platform [ ]. We also have extended the experimental setup in [ ] to include more pragmatic test inputs regarding motion graphs, i.e., graphs which model the topology of the motion of the hosts. Our test inputs included both random as well as more structured graphs. In conducting our experimental study, we were interested in providing measures on communication times (especially average message delay), message delivery rate, and support utilization (total number of messages contained in all members of the support).

Our experiments showed that: (i) for both protocols only a small support is required for efficient communication; (ii) the runners protocol outperforms the snake protocol in almost all types of inputs we considered. More precisely, the runners protocol achieve a better average message delay in all test inputs considered, except for the case of random graphs with a small support size. The new protocol achieves a higher delivery rate of messages right from the beginning, while the snake protocol requires some period of time until its delivery rate stabilizes to a value that is always smaller than that of runners. Finally, the runners protocol has smaller requirements for the size of local memory per member of the support.

Previous Work: A similar approach is presented in [ ] where a *compulsory* protocol is introduced in the sense that all users are forced to perform concurrent and independent random walks on the area covered by the network.

A model similar to the above is presented in [ ]. The protocol forces all mobile users to slightly deviate (for a short period of time) from their predefined, deterministic routes, in order to propagate the messages. This protocol is also compulsory for any host and it works only for deterministic host routes.

Adler and Scheideler [ ] in a previous work, dealt only with *static* transmission graphs, i.e., the situation where the positions of the mobile hosts and the environment do not change. In [ ] the authors pointed out that static graphs provide a starting point for the dynamic case. In our work, we consider the *dynamic case* (i.e., mobile hosts move *arbitrarily*) and in this sense we extend their work.

2   The Model of the Space of Motions

We use the graph theoretic model introduced in [ ] and also used in [ ] that maps the movement of the mobile users in the three-dimensional space S to a so-called *motion graph* $G = (V, E)$ (its precise definition will be given shortly). Let $|V| = n$. The environment where the stations move (in three-dimensional space with possible obstacles) is abstracted by a graph by neglecting the detailed geometric characteristics of the motion. We first assume that each mobile host has a transmission range represented by a sphere tr having the mobile host as its center. This means that any other host inside tr can receive any message broadcasted by this host. We approximate this sphere by a cube tc with volume $\mathcal{V}(tc)$, where $\mathcal{V}(tc) < \mathcal{V}(tr)$. The size of tc can be chosen in such a way that its volume $\mathcal{V}(tc)$ is the maximum integral value that preserves $\mathcal{V}(tc) < \mathcal{V}(tr)$, and if a mobile host inside tc broadcasts a message, this message is received by any other host in tc. Given that the mobile hosts are moving in the space S, S is quantized into cubes of volume $\mathcal{V}(tc)$.

The *motion graph* $G = (V, E)$, which corresponds to the above quantization of S, is constructed in the following way. There is a vertex $u \in V$ representing a cube of volume $\mathcal{V}(tc)$. Two vertices u and v are connected by an edge $(u, v) \in E$ if the corresponding cubes are adjacent. Let $V(S)$ be the volume of space S. Clearly, $n = V(S)/V(tc)$ and consequently n is a rough approximation of the ratio $V(S)/V(tr)$.

3   Description of the Implemented Protocols

We start with a few definitions that will be used in the description of the protocols. The subset of the mobile hosts of an ad-hoc mobile network whose motion is determined by a network protocol P is called the *support* Σ of P. The part of P which indicates the way in which the members of Σ move and communicate is called the *support management subprotocol* M_Σ of P. In addition, we may wish that the way hosts in Σ move and communicate can tolerate failures of hosts. In such a case, the protocol is called *robust*. A protocol is called *reliable* if it allows the sender to be notified about delivery of the information to the receiver.

We assume that the motions of the mobile users which are not members of Σ are arbitrary but *independent* of the motion of the support (i.e., we exclude the case where some of the users not in Σ are deliberately trying to avoid Σ). This is a pragmatic assumption usually followed by application protocols. In the following, we assume that message exchange between nodes within communication distance of each other takes negligible time (i.e., the messages are short packets and the wireless transmission is very fast). Following [,], we further assume that all users (even those not in the support) perform independent and concurrent random walks.

3.1   The Snake Protocol

The main idea of the protocol proposed in [ ] is as follows. There is a set-up phase of the ad-hoc network, during which a predefined number, k, of hosts, become the nodes of the support. The members of the support perform a leader election, which is run once and imposes only an initial communication cost. The elected leader, denoted by MS_0, is used to co-ordinate the support topology and movement. Additionally, the leader assigns local names to the rest of the support members $MS_1, MS_2, ..., MS_{k-1}$.

The nodes of the support move in a coordinated way, always remaining pairwise adjacent (i.e., forming a list of k nodes), so that they sweep (given some time) the entire motion graph. Their motion is accomplished in a distributed way via a *support motion subprotocol* P_1. Essentially the motion subprotocol P_1 enforces the support to move as a "snake", with the head (the elected leader MS_0) doing a random walk on the motion graph and each of the other nodes MS_i executing the simple protocol "move where MS_{i-1} was before". When some node of the support is within the communication range of a sender, an underlying *sensor subprotocol* P_2 notifies the sender that it may send its message(s). The messages are then stored in every node of the support using a *synchronization subprotocol* P_3. When a receiver comes within the communication range of a node of the support, the receiver is notified that a message is "waiting" for him and the message is then forwarded to the receiver. Duplicate copies of the message are then removed from the other members of the support.

In this protocol, the support Σ plays the role of a (moving) backbone subnetwork (of a "fixed" structure, guaranteed by the motion subprotocol P_1), through which all communication is routed. The theoretical analysis in [ ] shows that the average message delay or communication time of the snake protocol is bounded above by the formula $\frac{2}{\lambda_2(G)} \Theta(n/k) + \Theta(k)$ where G is the motion graph, $\lambda_2(G)$ is its second eigenvalue, n is the number of vertices in motion graph G, and $k = |\Sigma|$. More details for the snake protocol can be found in [ ]. It can be also proved (see [ ]) that the snake protocol is reliable and partially robust (resilient to one fault).

3.2   The Runners Protocol

A different approach to implement M_Σ is to allow each member of Σ not to move in a snake-like fashion, but to perform an *independent* random walk on the motion graph G, i.e., the members of Σ can be viewed as "runners" running on G. In other words, instead of maintaining at all times pairwise adjacency between members of Σ, all hosts sweep the area by moving independently from each other. However, all communication is still routed through the support Σ. When two runners meet, they exchange any information given to them by senders encountered using a new synchronization subprotocol P_3'. As in the snake case, the underlying sensor sub-protocol P_2 notifies the sender that it may send its message(s). When a user comes within the communication range of a node of the support which has a message for the designated receiver $\mathcal{R}$, the waiting messages are forwarded to the receiver.

We expect that the size k of the support (i.e., the number of runners) will affect performance in a more efficient way than that of the snake approach. This expectation stems from the fact that each host will meet each other in parallel, accelerating the spread of information (i.e., messages to be delivered).

A member of the support needs to store all undelivered messages, and maintain a list of receipts to be given to the originating senders. For simplicity, we can assume a generic storage scheme where all undelivered messages are members of a set S_1 and the list of receipts is stored on another set S_2. In reality, the unique ID of a message and its sender ID is all that is needed to be stored in S_2.

When two runners meet at the same site of the motion graph G, the synchronization subprotocol P_3' is activated. The subprotocol imposes that when runners meet on the same site, their sets S_1 and S_2 are synchronized. In this way, a message delivered by some runner will be removed from the set S_1 of the rest of runners encountered, and similarly delivery receipts already given will be discarded from the set S_2 of the rest of runners. The synchronization subprotocol P_3' is partially based on the *two-phase commit* algorithm as presented in [ ] and works as follows.

Let the members of Σ residing on the same site (i.e., vertex) u of G be $MS_1^u, \ldots, MS_j^u$. Let also $S_1(i)$ (resp. $S_2(i)$) denote the S_1 (resp. S_2) set of runner MS_i^u, $1 \leq i \leq j$. The algorithm assumes that the underlying sensor sub-protocol P_2 informs all hosts about the runner with the lowest ID, i.e., the runner MS_1^u. P_3' consists of two rounds.

Round 1: All $MS_1^u, \ldots, MS_j^u$ residing on vertex u of G, send their S_1 and S_2 to runner MS_1^u. Runner MS_1^u collects all the sets and combines them with its own to compute its new sets S_1 and S_2: $S_2(1) = \bigcup_{1 \leq l \leq j} S_2(l)$ and $S_1(1) = \bigcup_{1 \leq l \leq j} S_1(l) - S_2(1)$.

Round 2: Runner MS_1^u broadcasts its decision to all the other support member hosts. All hosts that received the broadcast apply the same rules, as MS_1^u did, to join their S_1 and S_2 with the values received. Any host that receives a message at Round 2 and which has not participated in Round 1, accepts the value received in that message as if it had participated in Round 1.

This simple algorithm guarantees that mobile hosts which remain connected (i.e., are able to exchange messages) for two continuous rounds, will manage to synchronize their S_1 and S_2. Furthermore, on the event that a new host arrives or another disconnects during Round 2, the execution of the protocol will not be affected. In the case where runner MS_1^u fails to broadcast in Round 2 (either because of an internal failure or because it has left the site), then the protocol is simply re-executed among the remaining runners.

Remark that the algorithm described above does not offer a mechanism to remove message receipts from S_2; eventually the memory of the hosts will be exhausted. A simple approach to solve this problem is to construct, for each sender, an ordered list of message IDs contained in S_2. This ordered sequence of IDs will have gaps – some messages will still have to be delivered, and thus not part of S_2. In this list, we can identify the maximum ID before the first gap and remove from S_2 all message receipts with smaller ID. Although this is a rather simple approach, our experiments showed that it effectively reduced the memory usage.

The described support management subprotocol is clearly reliable. It is also robust as it is resilient to t faults, for any $0 \le t < k$. This can be achieved using redundancy: whenever two runners meet, they create copies of all messages in transit. In the worst-case, there are at most $k - t$ copies of each message. Note, however, that messages may have to be re-transmitted in the case that only one copy of them exists when some fault occurs.

4   Implementation and Experimental Results

4.1   Implementation Details

All of our implementations follow closely the support motion subprotocols described above. They have been implemented as C++ classes using several advanced data types of LEDA [ ]. Each class is installed in an environment that allows to read graphs from files, and to perform a network simulation for a given number of rounds, a fixed number of mobile users and certain communication and mobility behaviours. After the execution of the simulation, the environment stores the results again on files so that the measurements can be represented in a graphical way.

To extend our previous implementation [ ], we have redesigned the fundamental classes and their functionality in order to provide a generic framework for the implementation of any mobile protocol. Our implementation of ad-hoc mobile networks is now based on a new base class called mobile_host. This class is then extended to implement the class mh_user that models a user of the network that acts as sender and/or receiver, and the mh_snake and mh_runner classes that implement the support management subprotocols respectively. Also in our previous implementation we had a class called transmission_medium to realize the exchange of messages between mobile hosts. In the new implementation, this class has been replaced by a so-called environment class which handles not only message exchange, but also the movement of mobile hosts (which can follow various motion patterns).

It is worth noting that in [ ] we did not implement the synchronization sub-protocol P_3. This did not affect the behaviour of the protocol, but helped us avoiding to further complicate the implementation. However, in order to provide a fair comparison between the two different support management subprotocols (i.e., the snakes and runners), we have implemented in this work the subpro-tocol P_3 for the snake protocol, and we also counted the extra delay that was imposed by the synchronization of the mobile support hosts.

4.2   Experimental Setup

A number of experiments were carried out modeling the different possible sit-uations regarding the geographical area covered by an ad-hoc mobile network. We considered five kinds of inputs, unstructured (random) and more structured ones. Each kind of input corresponds to a different type of motion graph. We started with the graph families considered in [ ], namely random graphs, 2D grid graphs, and bipartite multi-stage graphs. We extended the above experimental setup by considering also 3D grid graphs and two-level motion graphs that are more close to pragmatic situations.

We considered several values for n in the range $[400, 6400]$ and different val-ues for the support size $k = |\Sigma|$, namely $k \in [3, 45]$. For each motion graph constructed, we injected $1,000$ users (mobile hosts) at random positions that generated 100 transaction message exchanges of 1 packet each by randomly pick-ing different destinations. Each experiment we carried out, proceeds in lockstep rounds called *simulation rounds* (i.e., we measure simulation time in rounds). Each mobile host h is considered identical in computing and communication capability. During a simulation round, h moves to an adjacent vertex of the mo-tion graph G and performs some computation. If $h \in \Sigma$, then its move and local computation are determined by the snake or the runners protocol. If $h \notin \Sigma$, then its move is random and with probability $p = 0.01$ the host h generates a new message (i.e., a new message is generated roughly every 100 rounds) by picking a random destination host. The selection of this value for p is based on the assumption that the mobile users do not execute a real-time application that requires continuous exchange of messages. Based on this choice of p, each exper-iment takes several thousands of rounds in order to get an acceptable average message delay. For each experiment, a total of $100,000$ messages were transmit-ted. We carried out each experiment until the $100,000$ messages were delivered to the designated receivers. Our test inputs are as follows.

Random Graphs. This class of graphs is a natural starting point in order to experiment on areas with obstacles. We used the $G_{n,p}$ model obtained by sampling the edges of a complete graph of n nodes independently with probabil-ity p. We used $p = \frac{1.05 \log n}{n}$ which is marginally above the connectivity threshold for $G_{n,p}$. The test cases included random graphs with $n \in \{1600, 3200, 6400\}$ over different values of $k \in [3, 45]$.

2D Grid Graphs. This class of graphs is the simplest model of motion one can consider (e.g., mobile hosts that move on a plane surface). We used two different $\sqrt{n} \times \sqrt{n}$ grid graphs with $n \in \{400, 1600\}$ over different values of k.

3D Grid Graphs. To evaluate the performance of our protocols over 3D space, we considered 3D grid graphs $(n^{1/3} \times n^{1/3} \times n^{1/3})$ to model the motion of hosts in 3D space. We used three different such graphs with $n \in \{512, 1000, 1280\}$ over different values of $k \in [3, 45]$.

Bipartite multi-stage graphs. A bipartite multi-stage graph is a graph consisting of a number of stages (or levels) ξ. Each stage contains n/ξ vertices and there are edges between vertices of consecutive stages. These edges are chosen randomly with some probability p among all possible edges between the two stages. This type of graphs is interesting as they model movements of hosts that have to pass through certain places or regions, and have a different second eigenvalue than grid and $G_{n,p}$ graphs (their second eigenvalue lies between that of grid and $G_{n,p}$ graphs). In our experiments, we considered $\xi = \log n$ stages and choose $p = \frac{\xi}{n} \log \frac{n}{\xi}$ which is the threshold value for bipartite connectivity (i.e., connectivity between each pair of stages). The test cases included multi-stage graphs with 7, 8 or 9 stages, number of vertices $n \in \{1600, 3200, 6400\}$, and different values of $k \in [3, 45]$.

Two-level motion graphs. Motivated by the fact that most mobile users usually travel along favourite routes (e.g., going from home to work and back) that usually comprise a small portion of the whole area covered by the network (e.g., urban highways, ring roads, metro lines etc.), and that in more congested areas there is a high volume of user traffic (e.g., city centers, airport hubs, tourist attractions, etc.), we have considered a two-level motion graph family to reflect this situation (this family has some similarities with the two-level semirandom graphs considered in [ ] for completely different problems and settings).

Let $d(u, v)$ denote the distance between vertices u and v in G, i.e., the length of a shortest path joining them. A *two-level graph* consists of subgraphs of the motion graph representing congested areas that are interconnected by a small number of paths representing the most favourite routes among them. A subgraph G_c is defined by randomly selecting a vertex u of G and then declare as vertices of G_c those vertices $v \in G$ with $d(u, v) \leq c$, where c is a constant denoting the diameter of the congested area. Let f be the number of subgraphs defined. The paths representing favourite routes are specified as follows: temporarily replace each G_c by a supervertex, find shortest paths among supervertices, and finally select those edges that belong to shortest paths. If we end up with more than αf paths, where $\alpha > 1$ is a constant, then we either arbitrarily select αf of them, or find a minimum spanning tree that spans the supervertices and use its edges as the selected ones. The majority of hosts is forced to move either within the subgraphs (congested areas) or along the paths connecting subgraphs (favourite routes).

We constructed a number of different data sets by tuning the various parameters, i.e., the diameter c of the congested areas and the number f of selected vertices (number of subgraphs representing congested areas). Note that for each message generated, the destination is chosen at random among all users of the network (i.e., the sender-receiver pairs are not fixed).

4.3 Experimental Results

We measured the total delay of messages exchanged between pairs of sender-receiver users. For each message generated, we calculated the overall delay (in terms of simulation rounds) until the message was finally transmitted to the receiver. We used these measurements to experimentally evaluate the performance of the two different support management subprotocols.

The reported experiments for the five different test inputs we considered are illustrated in Figure 1. In these graphics, we have included the results of two instances of the input graph w.r.t. n, namely a smaller and a larger value of n (similar results hold for other values of n). Each curve in Figure 1 is characterized by the name 'Px', where P refers to the protocol used (S for snake and R for runners) and x is a 3 or 4 digit number denoting the value of n.

The curves reported for the snake protocol confirm the theoretical analysis in [ ]. That is, the average message delay drops rather quickly when k is small, but after some threshold value stabilizes and becomes almost independent of the value of k. This observation applies to all test inputs we considered. We also note that this behaviour of the snake protocol is similar to the one reported in [ ], although we count here the extra delay time imposed by the synchronization subprotocol P_3.

Regarding the runners protocol, we firstly observe that its curve follows the same pattern with that of the snake protocol. Unfortunately, we don't have a theoretical analysis for the new protocol to see whether it behaves as should expected, but from the experimental evidence we suspect that it obeys a similar but tighter bound than that of the snake protocol. Our second observation is that the performance of the runners protocol is slightly better than that of the snake protocol in random graphs (except for the case of random graphs with small support size; see Figure 1, top left) and bipartite multi-stage graphs (Figure 1, middle left), but is substantially better in the more structured cases (grid and two-level graphs; cf. Figure 1, top right and middle right). This could be partly explained by the fact that the structured graphs have a smaller second eigenvalue than that of bipartite multi-stage and random graphs. A small value for this eigenvalue makes the average message delay to be more dependent on the hidden constant in the asymptotic analysis of the snake protocol [ ] (see also Section 3.1) which is apparently larger than the corresponding constant of the runners protocol.

An interesting observation concerns the case of 2-level graphs, where the results look similar to those of grid graphs. Interestingly, for this case where the users do not perform random walks (as in the grid case) but their motion is getting more restricted (though still arbitrary), the performance characteristics of both protocols remain unchanged. Hence, we may conclude that for the case of more structured inputs the assumption of having the users performing continuous random walks over the motion graph G does not seem to affect performance. We believe that it is challenging to incorporate other sophisticated mechanisms to the support management subprotocols in order to take advantage of such user behaviour patterns.

In contrast to the experiments carried out in [ ], in this paper we also consider the utilization of the support protocol, i.e., the total number of multiple message copies stored in the support structure at any given time, as well as the message delivery rate.

In Figure 1 (bottom left) the total number of message copies generated by each protocol is shown for the first 7000 rounds of simulation on a bipartite multi-stage graph with $n = 6400$ and $k = 15$ (similar results hold for other test inputs and sizes). We observe that the snake protocol generates initially less copies than the runners. As the simulation goes on, the rate of generating redundant messages by the snake protocol increases faster than that of the runners. This can be partially justified by considering the average message delay. In the case of the snake, the overall message delay is worse than that of the runners, which implies that it will take longer to meet a sender. In addition, each message (and its copies) will remain for a longer period within the support structure of the snake until the designated receiver $\mathcal{R}$ is encountered. Therefore, at the beginning the snake protocol generates less copies, as less messages have been received, while as the simulation goes on, the total number of redundant copies increases as more messages are still pending. This observation implies that the runners protocol utilizes more efficiently the available resources as far as memory limitations are concerned.

Another way to evaluate the performance of the two protocols is to consider the message delivery rate. In Figure 1 (bottom right) the overall delivery rate of messages (in percentages) is projected over the simulation time (in rounds) for a 3D grid graph with $n = 6400$ and $k = 15$ (similar results hold for other test inputs and sizes). The snake protocol is slower in delivering messages than the runners protocol for the same reason explained above regarding utilization. As the simulation time increases, the two protocols finally reach high levels of delivery rates. However, it is clear that the runners will reach these high levels much faster compared to the snake. This provides an explanation regarding the total number of message copies stored within the support members of the snake protocol. Recall that the support synchronization subprotocol P_3 will need $O(k)$ time to generate k copies for each message, thus as the delivery rate of the snake is initially low, the total number of message copies will also be low. As the rate increases over time, P_3 will generate more message copies and thus increase the total number of message copies stored within the support structure.

5   Closing Remarks and Future Work

We have presented a new protocol for a basic communication problem in an ad-hoc mobile network and we provided a comparative experimental study of the new protocol (runners) and of a new implementation of an existing one (snake).

Our experiments showed that the new protocol outperforms the previous one in almost all test inputs we considered.

There are several directions for future work. An obvious one is to analyze theoretically the runners protocol. Another one is to enforce a specific motion pattern to the support (e.g., moving along a spanning subgraph of the motion graph) instead of doing random walks.

References

1. M. Adler and C. Scheideler. Efficient Communication Strategies for Ad-Hoc Wireless Networks. In *Proc. 10th Annual Symposium on Parallel Algorithms and Architectures* – SPAA'98, 1998. 160, 162
2. I. Chatzigiannakis, S. Nikoletseas, and P. Spirakis. Analysis and Experimental Evaluation of an Innovative and Efficient Routing Approach for Ad-hoc Mobile Networks. In *Proc. 4th Annual Workshop on Algorithmic Engineering* – WAE'00, 2000. 161, 162, 163, 165, 166, 168, 169
3. I. Chatzigiannakis, S. Nikoletseas, and P. Spirakis. An Efficient Routing Protocol for Hierarchical Ad-Hoc Mobile Networks. In *Proc. 1st International Workshop on Parallel and Distributed Computing Issues in Wireless Networks and Mobile Computing*, satellite workshop of IPDPS'01, 2001.
4. I. Chatzigiannakis, S. Nikoletseas, and P. Spirakis. Self-Organizing Ad-hoc Mobile Networks: The problem of end-to-end communication. To appear as a short paper In *Proc. 20th Annual Symposium on Principles of Distributed Computing* – PODC'01, 2001. 163
5. Z. J. Haas and M. R. Pearlman. The performance of a new routing protocol for the reconfigurable wireless networks. In *Proc. ICC'98*, 1998. 160, 163
6. K. P. Hatzis, G. P. Pentaris, P. G. Spirakis, V. T. Tampakas and R. B. Tan. Fundamental Control Algorithms in Mobile Networks. In *Proc. 11th Annual Symposium on Parallel Algorithms and Architectures* – SPAA'99, 1999. 162
7. G. Holland and N. Vaidya. Analysis of TCP Performance over Mobile Ad Hoc Networks. In *Proc. 5th Annual ACM/IEEE International Conference on Mobile Computing* – MOBICOM'99, 1999. 163
8. T. Imielinski and H. F. Korth. *Mobile Computing*. Kluwer Academic Publishers, 1996. 160
9. R. Iyer, D. Karger, H. Rahul, and M. Thorup. An experimental study of polylogarithmic fully-dynamic connectivity algorithms. In *Proc. 2nd Workshop on Algorithm Engineering and Experiments* – ALENEX 2000, pp. 59–78. 167
10. Q. Li and D. Rus. Sending Messages to Mobile Users in Disconnected Ad-hoc Wireless Networks. In *Proc. 6th Annual ACM/IEEE International Conference on Mobile Computing* – MOBICOM'00, 2000. 162
11. K. Mehlhorn and S. Näher. *LEDA: A Platform for Combinatorial and Geometric Computing*. Cambridge University Press, 1999. 161, 165
12. N. A. Lynch. *Distributed Algorithms*. Morgan Kaufmann Publishers Inc., 1996. 164
13. V. D. Park and M. S. Corson. Temporally-ordered routing algorithms (TORA): version 1 functional specification. IETF, Internet Draft, draft-ietf-manet-tora-spec-02.txt, Oct. 1999. 160
14. C. E. Perkins and E. M. Royer. Ad-hoc On demand Distance Vector (AODV) routing. IETF, Internet Draft, draft-ietf-manet-aodv-04.txt, 1999. 160

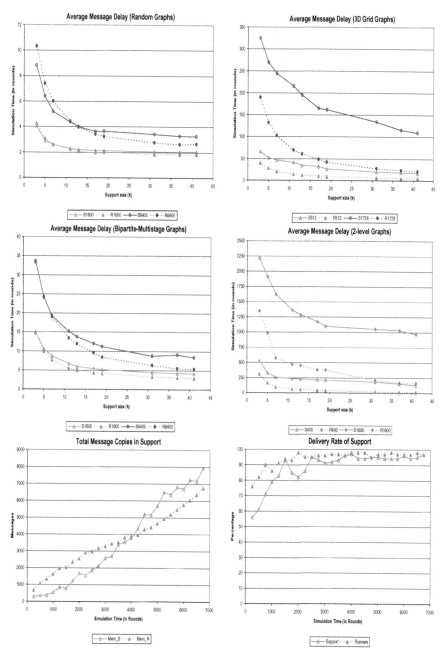

Fig. 1. Average message delay over various motion graphs (top and middle rows), total number of message copies (bottom left), and overall delivery rate of messages (bottom right)

Experimental Analysis of Algorithms for Bilateral-Contract Clearing Mechanisms Arising in Deregulated Power Industry

Chris Barrett[2], Doug Cook[1], Gregory Hicks[6], Vance Faber[4], Achla Marathe[2], Madhav Marathe[2], Aravind Srinivasan[3], Yoram J. Sussmann[5], and Heidi Thornquist[7]

[1] Department of Engineering, Colorado School of Mines
Golden CO 80401
[2] D-2 and CCS-3, Current address: P.O. Box 1663, MS M997
Los Alamos National Laboratory, Los Alamos NM 87545
{barrett,achla,marathe}@lanl.gov
[3] Lucent Technologies
600 Mountain Ave., Murray Hill, NJ 07974-0636[||]
srin@research.bell-labs.com
[4] Lizardtech Inc.
vance@lizardtech.com
[5] yorams@atl.fundtech.com
[6] Department of Mathematics, North Carolina State University
Raleigh
hicksgp@yahoo.com
[7] Department of Computational Science, Rice University
Houston, TX
heidi@caam.rice.edu

Abstract. We consider the bilateral contract satisfaction problem arising from electrical power networks due to the proposed deregulation of the electric utility industry in the USA. Given a network and a (multi)set of pairs of vertices (contracts) with associated demands, the goal is to find the maximum number of simultaneously satisfiable contracts. We study how four different algorithms perform in fairly realistic settings; we use an approximate electrical power network from Colorado. Our experiments show that three heuristics outperform a theoretically better algorithm. We also test the algorithms on four types of *scenarios* that are likely to occur in a deregulated marketplace. Our results show that the networks that are adequate in a regulated marketplace might be inadequate for satisfying all the bilateral contracts in a deregulated industry.

1 Introduction

The U.S. electric utility industry is undergoing major structural changes in an effort to make it more competitive [21,11,19,12]. One major consequence of the

---

[||] Part of this work was done while at the National University of Singapore

G. Brodal et al. (Eds.): WAE 2001, LNCS 2141, pp. 172–184, 2001.

deregulation will be to decouple the controllers of the network from the power producers, making it difficult to regulate the levels of power on the network; consumers as well as producers will eventually be able to negotiate prices to buy and sell electricity. See [ ]. In practice, deregulation is complicated by the facts that all power companies will have to share the same power network in the short term, with the network's capacity being just about sufficient to meet the current demand. To overcome these problems, most U.S. states have set up an ISO (independent system operator): a non-profit governing body to arbitrate the use of the network. The basic questions facing ISOs are how to decide which contracts to deny (due to capacity constraints), and who is to bear the costs accrued when contracts are denied. Several criteria/policies have been proposed and/or are being legislated by the states as possible guidelines for the ISO to select a maximum-sized subset of contracts that can be cleared simultaneously [ ]. These include: (a) Minimum Flow Denied: The ISO selects the subset of contracts that denies the *least* amount of proposed power flow. This proposal favors clearing bigger contracts first. (b) First-in First-out: The contract that comes first gets cleared first; this is the least discriminating to the contractors. (c) Maximum Consumers Served: This clears the smallest contracts first and favors the small buyers whose interests normally tend to go unheard.

There are three key issues in deciding policies that entail specific mechanisms for selecting a subset of contracts: *fairness* of a given policy to producers and consumers; the *computational complexity* of implementing a policy, and how *sound* a given policy is from an economic standpoint. (For instance, does the policy promote the optimal clearing price/network utilization etc.) Here we focus on evaluating the efficacy of a given policy with regard to its computational resource requirement and network resource utilization. It is intuitively clear that the underlying network, its capacity and topology, and the spatial locations of the bilateral contracts on the network, will play an important role in determining the efficacy of these policies. We do not discuss here the fairness and economics aspects of these policies: these are subjects of a companion paper. The work reported here was done as part of two inter-related projects at Los Alamos. The first project is to develop a mathematical and computational theory of simulations based on Sequential Dynamical Systems. The second aims at developing a comprehensive coupled simulation[1] of the *deregulated* electrical power industry [ ]. See http://www.lanl.gov/orgs/tsa/tsasa/ for additional details. To achieve this goal, we carry out the first experimental analysis of several algorithms for simultaneously clearing a maximal number of bilateral contracts. The algorithms were chosen according to provable performance, ability to serve as a proxy for some of the above-stated policies, and computational requirement. The algorithms are as follows; see § 3 for their specification. The ILP-RANDOMIZED ROUNDING (RR) algorithm has a provable performance guarantee under certain conditions. The computational resource requirement was quite high, but the approach also provides us with an upper bound on any optimal solution and proved useful in comparing the performance of the algorithms. The LARGEST-

---

[1] This coupled simulation has a market component and a physical network component.

FIRST HEURISTIC (LF) is a proxy for the *Minimum Flow Denied* policy. The SMALLEST-FIRST HEURISTIC (SF) serves as a proxy for the *Maximum Contracts Served* policy. The RANDOM-ORDER HEURISTIC (RO) clears the contracts in the random order. This algorithm was chosen as a proxy for the *First-in First-out* policy. Such a policy is probably the most natural clearing mechanism and is currently in place at many exchanges.

We used a coarse representation of the Colorado electrical power network (see § 4) to qualitatively compare the four algorithms discussed above in fairly realistic settings. The realistic networks differ from random networks and structured networks in the following ways: (i) Realistic networks typically have a very low average degree. In fact, in our case the average degree of the network is no more than 3. (ii) Realistic networks are not very uniform. One typically sees one or two large clusters (downtown and neighboring areas) and small clusters spread out throughout. (iii) For most empirical studies with random networks, the edge weights are chosen independently and uniformly at random from a given interval. However, realistic networks typically have very specific kinds of capacities since they are constructed with particular design goal.

From our preliminary analysis, it appears that although the simple heuristic algorithms do not have worst-case guarantees, they outperform the theoretically better randomized rounding algorithm. We tested the algorithms on four carefully chosen scenarios. Each scenario was designed to test the algorithms and the resulting solutions in a deregulated setting. The empirical results show that networks that are capable of satisfying all demand in a regulated marketplace can often be inadequate for satisfying all (or even a acceptable fraction) of the bilateral contracts in a deregulated market. Our results also confirm intuitive observations: e.g., the number of contracts satisfied crucially depends on the scenario and the algorithm.

As far as we are aware, this is the first study to investigate the efficacy of various policies for contract satisfaction in a deregulated power industry. Since it was done in fairly realistic settings, the qualitative results obtained here have implications for policy makers. To compare the algorithms in a quantitative and (semi-)rigorous way, we employ statistical tools and experimental designs. Although many of the basic tools are standard in statistics, the use of formal statistical methods in experimental algorithmics for analyzing/comparing the performance of heuristics has not been investigated to the best of our knowledge. We believe that such statistical methods should be investigated further by the experimental algorithmics community for deriving more quantitative conclusions when theoretical proofs are hard or not very insightful. Our results can also be applied in other settings, such as bandwidth-trading on the Internet. See, e.g., [ ]. Finally, to our knowledge, previous researchers have not considered the effect of the underlying network on the problems; this is an important parameter especially in a free-market scenario.

2   Problem Definitions

We briefly define the optimization problems studied here. We are given an undirected network (the power network) $G = (V, E)$ with capacities c_e for each edge e and a set of source-sink node pairs (s_i, t_i), $1 \leq i \leq k$. Each pair (s_i, t_i) has: (i) an integral *demand* reflecting the amount of power that s_i agrees to supply to t_i and (ii) a negotiated *cost* of sending unit commodity from s_i to t_i. As is traditional in the power literature, we will refer to the source-sink pairs along with the associated demands as *a set of contracts*. In general, a source or sink may have multiple associated contracts. We find the following notation convenient to describe the problems. Denote the set of nodes by N. The contracts are defined by a relation $R \subseteq (N \times N \times \Re \times \Re)$ so that tuple $(v, w, \alpha, \beta) \in R$ denotes a contract between source v and sink w for α units of commodity at a cost of β per unit of the commodity. For $A = (v, w, \alpha, \beta) \in R$ we denote $source(A) = v$, $sink(A) = w$, $flow(A) = \alpha$ and $cost(A) = \beta$. Corresponding to the power network, we construct a digraph $H = (V \cup S \cup T \cup \{s, t\}, E')$ with source s, sink node t, capacities $u : E' \to \Re$ and costs $c' : E' \to \Re$ as follows. For all $A \in R$, define new vertices v_A and w_A. Let $S = \{v_A \mid A \in R\}$ and $T = \{w_A \mid A \in R\}$. Each edge $\{x, y\}$ from G is present in H as the two arcs (x, y) and (y, x) that have the same capacity as $\{x, y\}$ has in G, and with cost 0. In addition, for all $A = (v, w, \alpha, \beta) \in R$, we introduce: (i) arcs (v_A, v) and (w, w_A) with infinite capacity and zero cost; (ii) arc (s, v_A) with capacity $flow(A) = \alpha$ and cost 0; and (iii) arc (w_A, t) with capacity $flow(A) = \alpha$ and cost equaling $cost(A)$. By this construction, we can assume without loss of generality that each node can participate in exactly one contract. A *flow* is simply an assignment of values to the edges in a graph, where the value of an edge is the amount of flow traveling on that edge. The value of the flow is defined as the amount of flow coming out of s (or equivalently the amount of flow coming in to t). A generic *feasible flow* $f = (f_{x,y} \geq 0 : (x, y) \in E')$ in H is any non-negative flow that: (a) respects the arc capacities, (b) has s as the only source of flow and t as the only sink. Note that for a given $A \in R$, in general it is not necessary that $f_{s,v_A} = f_{w_A,t}$. For a given contract $A \in R$, A is said to be *satisfied* if the feasible flow f in H has the additional property that for $A = (v, w, \alpha, \beta)$, $f_{s,v_A} = f_{w_A,t} = \alpha$; i.e., the contractual obligation of α units of commodity is shipped out of v and the same amount is received at w. Given a power network $G(V, E)$, a contract set R is *feasible* (or *satisfied*) if there exists a feasible flow f in the digraph H that satisfies every contract $A \in R$. Let $B = supply(s) = demand(t) = \sum_{A \in R} flow(A)$.

In practice, it is typically the case that R does not form a feasible set. As a result we have two possible alternative methods of relaxing the constraints: (i) relax the notion of feasibility of a contract and (ii) try and find a subset of contracts that are feasible. Combining these two alternatives we define the following types of "relaxed feasible" subsets of R. We will concern ourselves with only one variant here. A contract set $R' \subseteq R$ is a *0/1-contract satisfaction* feasible set if, $\forall A = (v, w, \alpha, \beta) \in R'$, $f_{s,v_A} = f_{w_A,t} = \alpha$.

Definition 1. *Given a graph $G(V, E)$ and a contract set R, the (0/1-VERSION, MAX-FEASIBLE FLOW) problem is to find a feasible flow f in H such that $\sum_{A \in R'} f(A)$ is maximized where R' forms a 0/1-contract satisfaction feasible set of contracts. In the related (0/1-VERSION, MAX-#CONTRACTS) problem, we aim to find a feasible flow f in H such that $|R'|$ is maximized, where R' forms a 0/1-contract satisfaction feasible set of contracts.*

Though such electric flow problems have some similarities with those from other practical situations, there are many basic differences such as reliability, non-distinguishability between the power produced by different generators, short life-time due to inadequate storage, line effects etc. [ ]. The variants of flow problems related to power transmission studied here are intuitively harder than traditional multi-commodity flow problems, since we *cannot distinguish between* the flow "commodities" (power produced by different generators). As a result, current solution techniques used to solve single/multi-commodity flow problems are not directly applicable to the problems considered here.

3   Description and Discussion of Algorithms

We work on the (0/1-VERSION, MAX-#CONTRACTS) problem here. Let n and m respectively denote the number of vertices and edges in the network G. In [ ], it was shown that (0/1-VERSION, MAX-#CONTRACTS) is NP-hard; also, unless $NP \subseteq ZPP$, it cannot be approximated to within a factor of $m^{1/2-\epsilon}$ for any fixed $\epsilon > 0$, in polynomial time. Thus, we need to consider good heuristics/approximation algorithms. First, there are three simple heuristics. The SMALLEST-FIRST HEURISTIC considers the contracts in non-decreasing order of their demands. When a contract is considered, we accept it if it can be feasibly added to the current set of chosen contracts, and reject it otherwise. The LARGEST-FIRST HEURISTIC is the same, except that the contracts are ordered in non-increasing order of demands. In the RANDOM-ORDER HEURISTIC, the contracts are considered in a random order.

We next briefly discuss an approximation algorithm of [ ]. This has proven performance only when all source vertices s_i are the same; however, this algorithm extends naturally to the multi-source case which we work on. An integer linear programming (ILP) formulation for the problem is considered in [ ]. We have indicator variables x_i for the contract between s_i and t_i, and variables $z_{i,e}$ for each (s_i, t_i) pair and each edge e. The intended meaning of x_i is that the total flow for (s_i, t_i) is $d_i x_i$; the meaning of $z_{i,e}$ is that the flow due to the contract between (s_i, t_i) on edge e is $z_{i,e}$. We write the obvious flow and capacity constraints. Crucially, we also add the valid constraint $z_{i,e} \leq c_e x_i$ for all i and e. In the integral version of the problem, we will have $x_i \in \{0, 1\}$, and the $z_{i,e}$ as non-negative reals. We relax the condition "$x_i \in \{0, 1\}$" to "$x_i \in [0, 1]$" and solve the resultant LP; let y^* be the LP's optimal objective function value. We perform the following rounding steps using a carefully chosen parameter $\lambda > 1$. (a) Independently for each i, set a random variable Y_i to 1 with probability x_i/λ, and $Y_i := 0$ with probability $1 - x_i/\lambda$. (b) If $Y_i = 1$, we will choose to satisfy

$(1 - \epsilon)$ of (s_i, t_i)'s contract: for all $e \in E$, set $z_{i,e} := z_{i,e}(1 - \epsilon)/x_i$. (c) If $Y_i = 0$, we choose to have no flow for (s_i, t_i): i.e., we will reset all the $z_{i,e}$ to 0. A deterministic version of this result based on *pessimistic estimators*, is also provided in [ ]; see [ ] for further details.

Theorem 1. ([ ]) *Given a network G and a contract set R, we can find an approximation algorithm for (0/1-*VERSION*, MAX-#*CONTRACTS*) when all source vertices are the same, as follows. Let OPT be the optimum value of the problem, and m be the number of edges in G. Then, for any given $\epsilon > 0$, we can in polynomial time find a subset of contracts R' with total weight $\Omega(OPT \cdot \min\{(OPT/m)^{(1-\epsilon)/\epsilon}, 1\})$ such that for all $i \in R'$, the flow is at least $(1 - \epsilon)d_i$.*

4   Experimental Setup and Methodology

To test our algorithms experimentally, we used a network corresponding to a subset of a real power network along with contracts that we generated using different scenarios. The network we used (as shown in Figure 1) is based on the power grid in Colorado and was derived from data obtained from PSCo's (Public Service Company of Colorado) Draft Integrated Resources Plan listing of power stations and major substations. We restricted our attention to major trunks only. The plant capacities were derived from the PSCo data. To generate maximum aggregate demands for the sink nodes we used data from the US Census bureau on the number of household per county. The demands were generated by assigning counties to specific sink nodes in the network.

All the test cases were generated from the basic model. The general approach we used was to fix the edge capacities and generate source-sink contracts combinations using the capacities and aggregate demands in the basic model as upper bounds. To ensure that the test cases we generated corresponded to (1) difficult problems, i.e. infeasible sets of contracts, and (2) problems that might reasonably arise in reality, we developed several scenarios that included an element of randomness.

The current implementation is still based upon a network which should be feasible only if the total source capacity is greater than the total sink capacity and the only requirement is that the total sink capacity be satisfied regardless of which source provides the power. **Scenarios 1, 2, 3** and **4** are based around the network with total generating capacity, **6249 MW**, and reduced sink capacities near **4400MW** combined.

Scenario 1: The source capacity of the network was reduced until the maximum flow in the network was slightly less than the demand.

Scenario 2: For this scenario, we took the basic network and increased the sink capacity while the source capacity remained fixed.

Scenario 3: The edge capacities were adjusted, reduced in most cases, to limit the network to a maximum flow of slightly more than 4400MW given its source

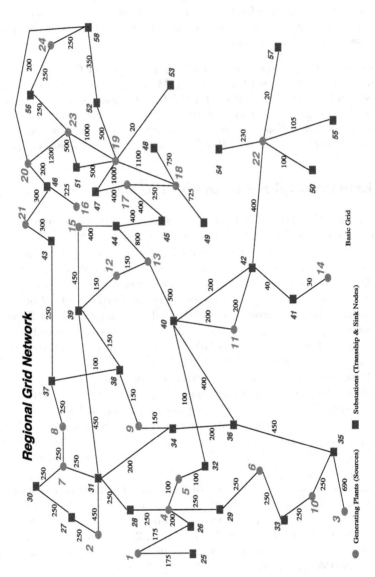

Fig. 1. This shows the network with nodes numbered as they are referenced in all scenarios and edge capacities labeled at values used for Scenarios 1 & 2. The placement of the nodes and edges are what is probably the final form. The least number of edges cross and the nodes in the upper right are spread out a little bit maintaining the general feel of the distribution while allowing easier reading

and sink distribution. Here, if the load is allowed to be fulfilled from any source (as is normally done with centralized control), the network and the edge capacities are enough to handle a total of 4400MW. However, if we insist that a particular source needs to serve a particular sink (as is done in bilateral contract satisfaction), then the capacities may not be enough to handle the same load of 4400MW.

Scenario 4: For this scenario, we took the network of Scenario 3 and biased the selection of source nodes towards the lower valued source units.

Methodology: We worked with the four scenarios and ran all four algorithms for each. For the three greedy heuristics the implementations are fairly straightforward, and we used public-domain network flow codes. Implementing the randomized rounding procedure requires extra care. The pessimistic estimator approach of [ ] works with very low probabilities, and requires significant, repeated re-scaling in practice. Thus we focus on the randomized version of the algorithm of [ ]; five representative values of ϵ varying from .1 to .5 were chosen. We believe that satisfying a contract partially so that a contract is assigned less than .5 of the required demand is not very realistic. For each scenario, and for each of the 5 values of ϵ, the programs implementing the algorithms under inspection produced 30 files from which the following information could be extracted: the running times and solutions of all four algorithms, and the LP upper bound on the ILP. The number 30 was chosen to ensure that a statistically "large" sample of each measure would be provided in order to make valid statistical inference. More attention is given to the quality-of-solution measure of an algorithm (rather than the running-time measure), since from a social standpoint, contract satisfaction may leave little room for finding solutions that are far from optimal. For a given algorithm $\mathcal{A}$, let $Value_\mathcal{A}$ denote the number of contracts that can be satisfied by $\mathcal{A}$. $p_\mathcal{A} = \frac{Value_\mathcal{A}}{\lfloor y^* \rfloor}$ provides a measure of the quality of the algorithm's solution. The value $\lfloor y^* \rfloor$ provides an upper bound on the objective function value. The objective of our experiments was to find out which, if any, of the algorithms discussed here performs better than the others, in terms of quality of solution and running time for different contract scenarios.

5   Results and Analysis

5.1   Statistical Analysis

We use a statistical technique known as *analysis of variance* (ANOVA) to test whether differences in the sample means of algorithms and scenarios reflect differences in the means of the statistical populations that they came from or are just sampling fluctuations. This will help us identify which algorithm and scenarios perform the best. See ([ ], [ ]) for details.

Quality of Solution: We first describe the experiment for the quality of solution, i.e., $p_\mathcal{A}$. We use a two-factor ANOVA model since our experiment involves two factors: the algorithms $\mathcal{A}_i$, $i = 1, 2, 3$ and 4, and scenarios $\mathcal{S}_j$, $j = 1, 2, 3$

and 4. Following classical statistics terminology, we will sometimes refer to algorithms as *treatments* and the scenarios as *blocks*. We will use $\mathcal{A}$ to denote the set of algorithms and $\mathcal{S}$ to denote the set of scenarios. For each algorithm-scenario pair we have *30 observations (or replicates)*. When testing the efficacy of the algorithms, we use 4 algorithms, each having 120 observations (30 for each scenario) from the corresponding population. The design of experiment used here is a *fixed-effect complete randomized block*. The model for randomized block design includes constants for measuring the scenario effect (block effect), the algorithm effect (treatment effect) and a possible interaction between the scenarios and the algorithms. An appropriate mathematical model for testing the above hypothesis is given by: $X_{ijk} = \mu + \tau_i + \beta_j + (\tau\beta)_{ij} + \varepsilon_{ijk}$, where X_{ijk} is the measurement (p_A) for the *kth* sample within the *ith* algorithm and the *jth* scenario. τ_i is the algorithm effect. β_j is the scenario effect. $(\tau\beta)_{ij}$ captures the interaction present between the algorithms and the scenarios. ε_{ijk} is the random error. We use S-Plus [16] software to run two-factor ANOVA to test the following three different hypotheses; as is standard, when discussing any one of these hypotheses, we refer to it as the *null hypothesis*.

1. Are the means given by the 4 different algorithms equal? That is, the hypothesis H_0 here is that $\tau_1 = \tau_2 = \tau_3 = \tau_4$.
2. Are the means given by the 4 different scenarios equal? The hypothesis H_0 here is that $\beta_1 = \beta_2 = \beta_3 = \beta_4$.
3. Is there any interaction between the two factors? H_0 here is that $(\tau\beta)_{ij} = 0$.

The results of two-factor ANOVA are shown in Table 1 and Table 2. In Table 1, the term $\overline{X}_{11\cdot}$, for instance, means that we fix $i = 1$ and $j = 1$, and take the average of X_{ijk} over all k. Similarly for other such terms. In the following discussion, we explain the meaning of each column. DF refers to the degrees of freedom, SS refers to the sum of squared deviations from the mean. MS refers to the mean square error, which is the sum of squares divided by the degrees of freedom. The p-value gives the smallest level of significance at which the null hypothesis can be rejected. The lower the p-value, the lesser the agreement between the data and the null hypothesis. Finally the F-**test** is as follows. To test the null hypothesis that the population means are equal, ANOVA compares two estimates of σ^2. The first estimate is based on the variability of each population mean around the grand mean. The second is based on the variability of the observations in each population around the mean of that population. If the null hypothesis is true, the two estimates of σ^2 should be essentially the same. Otherwise, if the populations have different means, the variability of the population mean around the grand mean will be much higher than the variability within the population. The null hypothesis in the F-test will be accepted if the two estimates of σ^2 are almost equal.

In a two-factor fixed-effect ANOVA, three separate F-**tests** are performed: two tests for the factors and the third for the interaction term. If the F-ratio is close to 1, the null hypothesis is true. If it is considerably larger–implying that the variance between means is larger than the variance within a population–the null hypothesis is rejected. The F-distribution table should be checked to see if the F-ratio is significantly large. The results in Table 2 show that all the above three null hypothesis are rejected at any significance level. This implies that the performance (measured by p_A) of at least one of the algorithms is significantly different from the other algorithms. Also, different scenarios make a difference in the performance. Finally, the scenarios and the algorithms interact in a significant way. The interaction implies that the performance of the algorithms are different for different scenarios. More details on what caused the rejection of null hypotheses and a discussion of running times are omitted here due to the lack of space. Interested readers can request a copy of the more complete version of this work by contacting one of the authors.

Table 1. The Mean Values of the Quality of Solution

Performance Measure: Quality of Solution (in %)				
	RR	SF	LF	RO	Scenario Means
Scenario 1	$\overline{X}_{11.}=48.68$	$\overline{X}_{21.}=99.73$	$\overline{X}_{31.}=97.97$	$\overline{X}_{41.}=97.78$	$\overline{X}_{.1.}=86.02$
Scenario 2	$\overline{X}_{12.}=46.91$	$\overline{X}_{22.}=99.56$	$\overline{X}_{32.}=98.38$	$\overline{X}_{42.}=98.93$	$\overline{X}_{.2.}=85.94$
Scenario 3	$\overline{X}_{13.}=45.69$	$\overline{X}_{23.}=99.25$	$\overline{X}_{33.}=97.10$	$\overline{X}_{43.}=98.82$	$\overline{X}_{.3.}=85.22$
Scenario 4	$\overline{X}_{14.}=46.99$	$\overline{X}_{24.}=98.03$	$\overline{X}_{34.}=88.65$	$\overline{X}_{44.}=93.41$	$\overline{X}_{.4.}=81.77$
Algo. Means	$\overline{X}_{1..}=47.07$	$\overline{X}_{2..}=99.14$	$\overline{X}_{3..}=95.51$	$\overline{X}_{4..}=97.24$	$\overline{X}_{...}=84.74$

Table 2. Results of Two-Factor ANOVA for Quality of Solution

Source	DF	SS	MS	F-test	p-value
Scenario (Block)	3	0.14	0.05	43.38	0
Algorithm (Treatment)	3	22.78	7.59	6792.60	0
Scenario:Algorithm	9	0.12	0.01	15.90	0
Residuals	464	0.40	.0008		
Total	479	23.45			

5.2   General Conclusions

1. Although there exist instances where the three heuristics produce solutions as large as $\Omega(n)$ times the optimal fractional solution, most of our tests show that we could find integral solutions fairly close to optimal.

2. Our experiments show that different scenarios make a *significant* difference in the type of solutions obtained. For example, the quality of solution obtained using the fourth scenario is significantly worse than the first three scenarios. The sensitivity to the scenarios poses interesting questions for infrastructure investment. The market will have to decide the cost that needs to be paid for expecting the necessary quality of service. It also brings forth the equity-benefit question: namely, who should pay for the infrastructure improvements?

3. It is possible that for certain scenarios, the underlying network is incapable of supporting even an acceptable fraction of the bilateral contracts. This observation although fairly intuitive provides an extremely important message, namely, networks that were adequate to service customers in a completely regulated power market *might* not be adequate in deregulated markets. This makes the question of evicting the bilateral contracts more important.

4. One expects a trade-off between the number of contracts satisfied and the value of ϵ, for the randomized rounding algorithm: as ϵ increases, and the demand condition is more relaxed, a higher number of contracts should get satisfied. But our experiments show that the change in the number of contracts satisfied for different values of ϵ is insignificant. Also, $\lambda = 2$ gave the best solutions in our experiments.

5. In practical situations, the Random-Order heuristic would be the best to use since it performs very close to the optimal in terms of quality of solution and has very low running time. Furthermore, though the Smallest-First heuristic does even better on many of our experiments, Random-Order is a natural proxy to model contracts arriving in an unforeseen way. Also, since the heuristics deliver solutions very close to the LP upper bound, we see that this LP bound is tight and useful. To further evaluate the randomized rounding algorithm, we need to implement its deterministic version [ ], which is a non-trivial task.

Acknowledgments: All authors except Aravind Srinivasan were supported by the Department of Energy under Contract W-7405-ENG-36. We thank Darrell Morgensen and Andy White for their support. Work by Doug Cook, Vance Faber, Gregory Hicks, Yoram Sussmann and Heidi Thornquist was done when they were visiting Los Alamos National Laboratory. We thank Dr. Martin Wildberger (EPRI) for introducing us to the problems considered herein, and for the extensive discussions. We thank Mike Fugate, Dean Risenger (Los Alamos National Laboratory) and Sean Hallgren (U.C. Berkeley) for helping us with the experimental evaluation. We also thank Massoud Amin (EPRI), Terence Kelly (University of Michigan), R. Ravi (CMU) and the members of the Power Simulation project (ElectriSim) at Los Alamos National Laboratory: Dale Henderson, Jonathan Dowell, Martin Drozda, Verne Loose, Doug Roberts, Frank Wolak (Stanford University) and Gerald Sheble (Iowa State University) for fruitful discussions. Finally, thanks to the WAE 2001 referees for their helpful suggestions.

References

1. Ahuja, R. K., Magnanti, T. L., Orlin, J. B.: Network flows: theory, algorithms, and applications. Prentice Hall, Englewood Cliffs, New Jersey. (1993)
2. Example sites for information on bandwidth trading: `http://www.williamscommunications.com/prodserv/network/webcasts/`, `http://www.lightrade.com/dowjones2.html bandwidth.html` 174
3. Barrett, C., Hunt, H. B., Marathe, M., Ravi, S. S., Rosenkrantz, D., Stearns, R.: Dichotomy Results for Sequential Dynamical Systems. To appear in Proc. MFCS (2001)
4. Cardell, J. B., Hitt, C. C., Hogan, W. W.: Market Power and Strategic Interaction in Electricity Networks. Resource and Energy Economics Vol. 19 (1997) 109-137.
5. California Public Utilities Commission. Order Instituting Rulemaking on the Commission's Proposed Policies Governing Restructuring California's Electric Services Industry and Reforming Regulation, Decision 95-12-063 (1995)
6. Cook, D., Faber, V., Marathe, M., Srinivasan, A., Sussmann, Y. J.: Low Bandwidth Routing and Electrical Power Networks. Proc. 25th International Colloquium on Automata, Languages and Programming (ICALP) Aalborg, Denmark, LNCS 1443, Springer Verlag, (1998) 604-615 176, 177, 179, 182
7. Dowell, L. J., Drozda, M., Henderson, D. B., Loose, V., Marathe, M., Roberts, D.: ELISIMS: Comprehensive Detailed Simulation of the Electric Power Industry. Technical Report LA-UR-98-1739, Los Alamos National Laboratory (1998) 173
8. The Changing Structure of the Electric Power Industry: Selected Issues. DOE 0562(98), Energy Information Administration, US Department of Energy, Washington, D. C. (1998)
9. Fryer, H. C.: Concepts and Methods of Experimental Statistics. Allyn and Bacon Inc. (1968) 179
10. Glass, G. V., Hopkins, K. D.: 1996. Statistical Methods in Education and Psychology. third edition (1996) 179
11. EPRI-Workshop: Underlying Technical Issues in Electricity Deregulation. Technical report forthcoming, Electric Power Research Institute (EPRI), April 25-27 (1997) 172
12. The U. S. Federal Energy Regulatory Commission. Notice of Proposed Rulemaking ("NOPRA"). Dockets No. RM95-8-000, RM94-7-001, RM95-9-000, RM96-11-000, April 24 (1996) 172
13. Garey, M. R., Johnson, D. S.: Computers and Intractability. A Guide to the Theory of NP-Completeness. Freeman, San Francisco CA (1979)
14. Hogan, W.: Contract networks for electric power transmission. J. Regulatory Economics. (1992) 211-242
15. Kolliopoulos, S. G., Stein, C.: Improved approximation algorithms for unsplittable flow problems. In Proc. IEEE Symposium on Foundations of Computer Science. (1997) 426-435
16. S-Plus5, "Guide to Statistics", MathSoft Inc. September (1988) 180
17. Srinivasan, A.: Approximation Algorithms via Randomized Rounding: A Survey. Lectures on Approximation and Randomized Algorithms (M. Karoński and H. J. Prömel, eds.), Series in Advanced Topics in Mathematics. Polish Scientific Publishers PWN, Warsaw.(1999) 9-71
18. Wildberger, A. M.: Issues associated with real time pricing. Unpublished Technical report, Electric Power Research Institute (EPRI) (1997) 173

19. Websites: http://www.magicnet.net/~metzler/page2d.html and http://
    www.enron.com/choice/dereg.fset.html (also see references therein).
    http://www.eia.doe.gov/ (Energy Information Administration) 172
20. Wolak, F.: An Empirical Analysis of the Impact of Hedge Con-
    tracts on Bidding Behavior in a Competitive Electricity Market,
    http://www.stanford.edu/~wolak. (2000)
21. The Wall Street Journal, August 04 2000, A2, August 14 2000 A6 and August
    21 2000, November 27 2000, December 13, 14 2000. 172
22. Wood, A. J., Wollenberg, B. F.: Power Generation, Operation and Control.
    John Wiley and Sons. (1996) 176

Pareto Shortest Paths is Often Feasible in Practice

Matthias Müller–Hannemann and Karsten Weihe

Forschungsinstitut für Diskrete Mathematik, Rheinische
Friedrich-Wilhelms-Universität Bonn
Lennéstr. 2, 53113 Bonn, Germany
{muellerh,weihe}@or.uni-bonn.de

Abstract. We study the problem of finding all Pareto–optimal solutions for the multi–criteria single–source shortest–path problem with nonnegative edge lengths. The standard approaches are generalizations of label-setting (Dijkstra) and label-correcting algorithms, in which the distance labels are multi–dimensional and more than one distance label is maintained for each node. The crucial parameter for the run time and space consumption is the total number of Pareto optima. In general, this value can be exponentially large in the input size. However, in various practical applications one can observe that the input data has certain characteristics, which may lead to a much smaller number — small enough to make the problem efficiently tractable from a practical viewpoint.

In this paper, we identify certain key characteristics, which occur in various applications. These key characteristics are evaluated on a concrete application scenario (computing the set of best train connections in view of travel time, fare, and number of train changes) and on a simplified randomized model, in which these characteristics occur in a very purist form. In the applied scenario, it will turn out that the number of Pareto optima on each visited node is restricted by a small constant. To counter–check the conjecture that these characteristics are the cause of these uniformly positive results, we will also report negative results from another application, in which these characteristics do not occur.

Keywords. Multi-criteria optimization, Pareto set, shortest paths, railway networks

1 Introduction

Problem and background. Multi–criteria optimization has been extensively studied due to its various applications in practical decision problems in operations research and other domains. The fundamental algorithmic problem is to find all *Pareto optima*. To define Pareto optimality, consider two d–dimensional real–valued vectors $x = (x_1, \ldots, x_d)$ and $y = (y_1, \ldots, y_d)$. If $x \neq y$ but $x_i \leq y_i$ for all $1 \leq i \leq d$ (in other words, if $x \leq y$ and $x_i < y_i$ for at least one i), then x *dominates* y, and we write $x < y$. Given a finite set X of d-dimensional vectors, we call $x \in X$ *Pareto–optimal* in X if there is no $y \in X$ that dominates x. Exact

G. Brodal et al. (Eds.): WAE 2001, LNCS 2141, pp. 185–197, 2001.
© Springer-Verlag Berlin Heidelberg 2001

multi–criteria optimization then means computing all Pareto–optimal elements of the solution space.

Multi–criteria shortest–path problems with nonnegative edge lengths is one of the most fundamental examples and occurs in many applications. To mention just a few, problems of this kind arise in communication networks (cost vs. reliability), in individual route planning for trucks and cars (fuel costs vs. time), in route guidance [ ], and in curve approximation [,]. Formally, we are given a directed graph $G = (V, E)$, a fixed positive integral number d, and d nonnegative lengths for each edge. Thus, each path in G has d lengths.

Like for the normal shortest–path problem ($d = 1$), it makes sense to consider the node–to–node case (paths from $s \in V$ to $t \in V$), the single–source case (paths from $s \in V$ to all other nodes), and the all–pairs case. In this paper, we will focus on the node–to–node and the single–source case. For the normal shortest–path problem, Dijkstra's algorithm is the standard approach for both cases. The difference is that, in the node–to–node case, the algorithm may terminate once the processing of t has been finished.

Tractability and state of the art. We sketch the previous work on multi–criteria shortest path problems only briefly. For a more complete overview, we refer to the section on shortest paths in the recent annotated bibliography on multi–objective combinatorial optimization [ ].

The standard approaches to the case that *all* Pareto optima have to be computed are generalizations of the standard algorithms for the single–criterion case. Instead of one scalar distance label, each node $v \in V$ is assigned a number of d–vectors, which are the lengths of all Pareto–optimal paths from s to v (clearly, for $d = 1$ the Pareto optima are exactly the distance labels). For the bicriteria case, generalizations of the standard label setting [ ] and label correcting [ ] methods have been developed. In the monograph of Theune [ ], algorithms for the multi-criteria case are described in detail in the general setting of cost structures over semi-rings. A *two-phase method* has been proposed by Mote et al. [ ]. They use a simplex-type algorithm to find a subset of all Pareto-optimal paths in the first phase, and a label-correcting method to find all remaining Pareto-optimal paths in the second phase.

The crucial parameter for the run time and the space consumption is the total number of Pareto optima over all visited nodes. It is well known that this number is exponential in $|V|$ in the worst case. This insight has been motivating two threads of research: theoretical research on approximation algorithms and on variations of the problem and experimental research on random inputs with specific characteristics.

Hansen [ ] and Warburton [ ] (the latter seemingly unaware of Hansen's work) both present a fully polynomial approximation scheme (FPAS) for finding a set of paths which are approximately Pareto–optimal in a certain sense. Moreover, Hansen [ ] considers ten different variations of bicriteria path problems and analyzes their computational complexity. In particular, he showed the exponential growth of the number of Pareto optima for the bicriteria shortest–path problem. The *(resource–)constrained* or *weight–restricted shortest–path problem* [ ] is a simplifying (yet still $\mathcal{NP}$–hard) variation of the bicriteria case. Here only one Pareto–optimal path is to be computed, namely the one that optimizes the first criterion subject to the condition that the second criterion does not exceed a given threshold value.

There are several experimental studies. Mote et al. [ ] investigate problem instances on random graphs and grid graphs with a positive correlation between the two length values of each arc. More precisely, the first length value is randomly generated from a uniform distribution within a certain range, whereas the second length value is a convex combination of the first length value and a randomly generated value from the same distribution. Their experiments indicate that the number of Pareto-optimal paths decreases with increasing correlation and that the overall number of such paths is quite small. Brumbaugh-Smith and Shier [ ] studied implementations of label-correcting algorithms on graphs where pairs of arc lengths are randomly generated from a bivariate normal distribution. For such instances, their empirical finding was that the asymptotic run time of the label-correcting method has a very good fit for $O(m\sqrt{p})$, where p denotes the average number of labels per node.

Contribution and overview. In various practical applications one can observe that, in contrast to the worst case, the total number of Pareto optima is small enough to make the problem efficiently tractable from a practical viewpoint. Generally speaking, the contribution of this paper is two–fold: we will identify certain key characteristics, which occur in various applications, and an experimental study will substantiate the claim that these characteristics may indeed be regarded as responsible for these good results.

This outcome may provide guidance for practical work in concrete applications. If the input data has these characteristics, it is not necessary to work on approximation algorithms or algorithms for the constrained version; the brute–force computation of the exact result is a feasible alternative. If the end–user of a system (e.g. a customer of a traffic information system) shall only be faced with a selection of Pareto–optimal paths, this selection can then be computed by a filter tool. The crucial point is this: if the algorithm delivers *all* Pareto optima to the filter tool, the selection may reflect the end–user's preferences, whereas otherwise it mainly reflects the details of the algorithm (which determine the selection delivered by the algorithm).

The first part of the experimental study is based on a concrete real–world scenario: computing the set of best train connections in view of travel time, fare, and number of train changes. The directed graph underlying this experiment is derived from the daily train schedule as follows. See Fig. 2. The arrival or departure of a train at a station will be called an *event*. The train graph contains one node for every event. Two events v and w are connected by an arc from v to w if v represents the departure of a train at some station and w represents the arrival of this train at the next station. Furthermore, two successive events at the same station are connected by an edge (in positive time direction; we consider the first event on each day as a successor of the last event before midnight).

Hence, in this graph, every station is represented as a cycle through all of its events.

This graph will be called the *train graph* in the following. See [ ] for details of this scenario and the application background. It will turn out that the number of Pareto optima for each visited node can be safely regarded as constant (Experiment 1 in Section 3), and that this result is quite robust against modifications of the scenario (Experiments 2–4).

This concrete application scenario shares a few fundamental characteristics with many other applications: the individual length functions are positively correlated (in a weak, heuristic, non–statistical sense), and the edges are partitioned into priority classes, which determine the rules of this correlation. Moreover, we will consider a typical characteristic of the output paths, which we will call *restricted non–monotonous* or *bitonic* (Experiment 5). The impact of all of these characteristics is investigated in the second part of the experimental study. There we will consider a simplified, randomized model, in which these two characteristics occur in a quite purist form. The results suggest that the presence of these characteristics indeed reduces the number of Pareto optima to a practically tractable number. To support the conjecture that these characteristics are the cause of these uniformly positive results, we will also report negative results from another application, in which these characteristics do not occur (Experiment 6).

The paper is organized as follows. In Section 2 we will introduce a formal model for the above–mentioned characteristics of the length functions. We analyse this model and present positive and negative worst–case results. Section 3 contains selected results of our experimental study in public railroad traffic, whereas Section 4 contains selected results of the experimental study on artificial instances.

The main focus of our study is the number of Pareto optima, not the run time of some piece of code. As mentioned above, this parameter decides upon tractability and non–tractability, no matter how fine–tuned the code is. Our code was tuned to compute this parameter and some related data. It does not compute paths. Hence, it does not make sense to present rough CPU times.

2   Formal Models of Characteristics

In the general *edge–class model*, we regard the edge set E of the underlying graph as partitioned into $k \geq 1$ classes, $E_1, E_2, \dots, E_k$.

Examples:

1. In the above–mentioned train graph, each edge that represents a train move from one stop to the next one inherits its class from the class of the train (high–speed trains, fast trains, local trains, etc.), and the other edges form a class of its own. In many countries (incl. Germany), the fare regulations are different for the individual train classes.

Fig. 1. Bad examples with only two different cost ratios

2. In route planning, the nodes of the graph are road crossings and the like, and the edges are segments of roads between successive crossings. Each edge belongs to a highway, national road, local road, etc.
3. In communication networks, we have central backbone edges, local backbone edges, private connections, etc.

The basis of the second part of the study will be a fundamental special case of the general edge–class model, which we call the *ratio–restricted lengths model*.

Ratio–restricted lengths model. In this special case, we only have two length functions (bicriteria case). Each class E_i is equipped with a value $r_i > 0$, where $r_1 < r_2 < \cdots < r_k$. For each edge $e \in E_i$, the ratio between the first and the second length value of e equals r_i. Observe that the problem does not change if all ratios are multiplied by a common strictly positive factor. Hence, we may assume $r_1 = 1$ without loss of generality.

For example, the ratio–restricted lengths model makes some sense in case of route planning. There is a restricted number of speed limits, and each speed gives a constant ratio between the travel time and the distance. Whenever the speed is fixed, the ratio between the distance and the required amount of fuel can also be regarded as constant. Thus, the ratio between time and fuel cost is constant for all road segments with the same speed limit.

In theory, the ratio–restricted lengths model is not significantly better than the general case:

Lemma 1. *Even in the ratio–restricted lengths model with at most $k > 1$ different ratios, the number of Pareto-optima can be exponentially large.*

Proof. We construct such an instance, which is basically an acyclic chain, with two alternative paths between node v_{2i} and v_{2i+2}, see Fig. 1. Formally, let $G = (V, E)$ be a digraph with $n+1$ nodes, numbered from 0 to n. The arc set consists of the arcs $v_{2i} \rightarrow v_{2i+1}$ and $v_{2i+1} \rightarrow v_{2i+2}$ with lengths $(2^{i+1}, 2^i)$, and the arcs $v_{2i} \rightarrow v_{2i+2}$ with lengths $(2^{i+1}, 2^{i+2})$. Hence, only the two different ratios $\frac{1}{2}$ and 2 occur.

Now it is easy to prove by induction, that there are 2^i different Pareto-optimal paths at node $2i$. The objective values of the Pareto-optimal labels at node $2i$ are of the form $(2^i - 2 + 2j, 2^{i+1} - 4 - 2j)$ for $j = 0, 1, \ldots, 2^i - 1$ and $i \geq 1$. □

In practice, one can often observe that the best paths are *restricted non–monotonous* in the following sense. For an edge $e \in E$ in the edge–class model, let $class(e) = i$ for $e \in E_i$. Let $e_1 - e_2 - \cdots - e_m$ be the edges of a path in the order in which they occur on the path. This path is *up–monotonous*, if

$class(e_i) < class(e_j)$ for $i < j$ and *down–monotonous*, if $class(e_i) > class(e_j)$
for $i < j$. We can decompose a path into its inclusion–maximal up–monotonous
and down–monotonous sections. We will call a path *restricted non–monotonous*
with restriction ℓ if neither the number of up–monotonous segments nor the
number of down–monotonous segments exceeds ℓ. In case of $\ell = 1$, we will speak
of *bitonic* paths.

In practice, the best paths are often even bitonic. For example, it is quite
untypical that an optimal overland car route includes a local road between two
sections on highways, or that an optimal train connection includes a local train
between two high–speed trains. There are exceptions: for example, in Paris one
has to use the subway from one station to another station to change high–speed
trains.

This observation gives rise to two questions: how good are optimal bitonic
paths compared to the overall optimal paths, and does the restriction to bitonic
paths significantly reduce the total number of Pareto optima on the visited
nodes. This question is important in practice, since some commercial tools pre-
fer or even restrict the search space to bitonic paths (statement from personal
communication).

In the first part of the computational study (in the concrete application
scenario), we will investigate these questions to some extent by checking the
computed Pareto–optimal paths for bitonicity.

In theory, even the restriction to bitonic paths does not pay off:

Lemma 2. *There is a class of instances where the number of Pareto–optimal
monotonous (bitonic) shortest paths is not bounded by a polynomial in $|V|$, if
the number of edge classes is not bounded by a constant.*

Proof. Omitted, due to space limitations.

If we combine the ratio–restricted lengths model with the restriction to restricted non–monotonous paths, we obtain a polynomial growth:

Lemma 3. *In the ratio-restricted lengths model with k distinct ratios, the num-
ber of Pareto–optimal monotonous or bitonic s-t-paths is bounded by $O(|V|^{k-1})$
or $O(|V|^{2k-2})$, respectively. This upper bound is tight.*

Proof. Omitted, due to space limitations.

This combination of two restrictions is the basis of the second part of the
study.

3   Pareto-Optimal Paths in Train Graphs

Recall the definition of the train graph from the introduction. Based on the given
data we can associate the following lengths functions to each edge:

1. travel distance, as the geometric distance between two stations;
2. travel time, as the time difference of the events representing the endnodes of an arc.

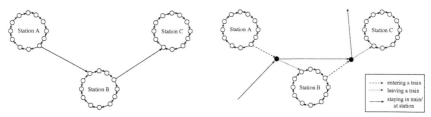

Fig. 2. Train graph model without (left) and with (right) a representation of distinguishable trains

This simple train graph model does not allow to capture the number of train changes for a connection between two stations as an optimization criterion. To make this possible, we enlarge the train graph model. Suppose that train T leaves station A at some event u, stops the next time at station B at event v_1 and continues its journey at event v_2 on its way to station C where it arrives at event w. Now we split the arcs $u \rightarrow v_1$ and $v_2 \rightarrow w$ from the train graph into chains of two arcs by introducing intermediate *train nodes* T_B and T_C, respectively. Finally we connect the train nodes by an arc $T_B \rightarrow T_C$. See Fig. 2. The arcs $u \rightarrow T_B$ and $v_2 \rightarrow T_C$ can now be interpreted as *entering* train T, the arcs $T_B \rightarrow v_1$ and $T_C \rightarrow w$ as *leaving* train T, and arc $T_B \rightarrow T_C$ means *staying on* the train T. Thus we get an additive cost counting the number of used trains for any path between two event nodes if we charge one cost unit for arcs representing entering a train, and zero units for all other arcs.

Our data is based on the public train schedule of the Deutsche Bahn AG in the winter period 1996/97 within Germany. In this case the train graph without train changes has a size of about 1 million nodes and roughly 1.4 million edges, the augmented model has roughly 1.4 million nodes and 2.3 million edges.

Fare model. From these data, we derive a third cost function, which is a model for travel fares and approximates the regulations of the Deutsche Bahn AG. In our model we assume that the basic fare is proportional to the travel distance. We distinguish between different *train classes*: high speed trains (ICE), fast long-distance trains (Intercity, Eurocity), ordinary long-distance trains (Interregio), regional commuter trains, and overnight trains. Depending on the train class we charge extra fares. For the fastest trains, the charge is proportional to the speed of the train (that is, from the point of view of a costumer, proportional to the time gain obtained for using a faster train).

More formally, our fare model is as follows. For an arc a between two different stations, we denote by $\ell(a)$ its length (the Euclidian distance between the two stations), and by $t(a)$ the travel time of the train associated with this arc. Then the fare $f(a)$ of this arc is computed by

$$f(a) := c_1 \ell(a) + c_2 \ell(a) \frac{\ell(a)}{t(a)} + c_3(a),$$

where c_1 is a constant price factor per kilometer, c_2 is a constant price factor for the speed, and $c_3(a)$ is a train class-dependent extra fee.

Experimental setup. In the following experiments we used the mentioned train graph. We used two different types of experiments:

Type 1: We performed bicriteria single–source shortest path computations from 1000 randomly selected source nodes and counted the number of different Pareto-optimal labels for each node.

Type 2: We used 1000 node–to–node customer queries stemming from a snapshot of the central *Hafas*[1] server of the Deutsche Bahn AG. To reduce the number of investigated labels, we used a simple heuristic improvement and discard all intermediate labels which are dominated by nodes at the target station. Its correctness follows from the additivity of our cost functions, the non-negativity of all cost coefficients, and the fact that subpaths of Pareto-optimal labels must also be Pareto-optimal. (Essentially the same trick has recently been used to speed up the single criterion Dijkstra algorithm for the one source many targets case [ ].)

Experiment 1. *We investigate the number of Pareto-optimal paths in the train graph for the two cost measures travel distance and travel time. The travel distance is given in kilometers, whereas time is measured in minutes.*

Observations. The number of labels stored at each node is really small, on average over all trials it is only 1.97, the maximum average number of labels in one trial is 3.69, and the overall maximum number of labels found at one node is 8. See the histogram in Fig. 3 for the average frequencies.

Experiment 2. *How sensitive are the results from Experiment 1 to the precision of the given distance measures?*

To study the robustness of the previous results against the precision of the edge lengths, we now rerun Experiment 1 with the travel distance between two stations given in a precision of one meter.

Observations and conclusion. The results are only slightly different, the number of labels stored at each node on average over all trials is now 2.00 compared with the 1.97, the maximum average number of labels in one trial is 3.66, but the overall maximum number of labels found at one node is still 8. Our observations indicate that the problem behaves rather insensitive to changes of the data precision.

---

[1] *Hafas* is a trademark of the Hacon Ingenieurgesellschaft mbH, Hannover, Germany

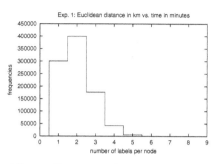
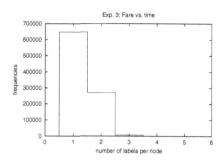

Fig. 3. Histogram for experiment 1   **Fig. 4.** Histogram for experiment 3

Experiment 3. *How many Pareto optima can be expected if we use travel fare vs. travel time as objectives, and how robust are these numbers against different fare models?*

We remodelled a simplified version of the current fare system of the Deutsche Bahn AG. Currently, the base fare is $c_1 = 0.27DM/km$ (for second class passengers). Using an Intercity/Eurocity train requires an extra charge of $c_3 = 7DM$. For the fastest train class (ICE) we use a speed dependent extra charge with a speed factor of $c_2 = 4\frac{DM*min}{km^2}$. This price system is in the following referred to as our Variant 1.

To study the robustness of our results we also considered several variations of the cost model.

- Variant 2: The same cost model as in Variant 1 but $c_1 = 0.41DM/km$ (the current base fare for first class passengers).
- Variant 3: The same cost model as in Variant 1 but $c_1 = 0.16DM/km$ (a cheap base fare).
- Variant 4: The same cost model as in Variant 1 but no additive charge for Intercity/Eurocity trains $(c_3 = 0)$. Instead, Intercity/Eurocity trains are treated like ICE trains and are charged in speed dependent fashion.
- Variant 5: The same cost model as in Variant 1 but the speed dependent factor for ICE trains is lowered to $c_2 = 1$.

For all variants we used the above-mentioned 1000 representative queries (Type 2).

Observations and conclusion. See Table 1 for an overview of the 5 variants. For all variants, the average numbers of Pareto labels per node is only .3, that is, very small. Due to the simple heuristic to discard labels which are dominated by labels at the destination only 27 percent of all nodes are visited on average to answer one query. The results are very insensitive to variations of the cost model.

Experiment 4. *What is the increase in the number of Pareto-optima if we use three optimization criteria in the train graph scenario?*

Table 1. Results for the different cost models in Experiment 3 and Experiment 5 (last row)

Variant	aver. # labels per node	aver. # labels at terminal	max. # labels at terminal	percentage of visited nodes
1	.302	3.36	21	27.20
2	.300	3.17	22	27.20
3	.303	3.51	21	27.20
4	.301	3.14	20	27.19
5	.299	3.01	21	27.16
tri-criteria	2.1	9.7	96	30.20

Table 2. Results for the elevation models in Experiment 6

data set	# of nodes	# of edges	max. # labels	aver. # labels
Austria small	11648	46160	661	253.15
Austria big	41600	165584	2543	994.07
Scotland small	16384	65024	955	288.66
Scotland big	63360	252432	2659	642.27

In our experiment, we used again the same queries and investigated the case of fare (in Variant 1) vs. travel time vs. train changes.

Observations and conclusion. See again Table 1. On average over the 1000 queries, the number of Pareto-optimal solutions found at the destination station was 9.7, the maximum was as large as 96. Over all nodes, the average number of labels was 2.1. The discarding heuristic still works well, the percentage of visited nodes increases slightly to 30.2 percent.

Experiment 5. *How many Pareto-optimal paths are bitonic in the train graph model?*

We rerun the previous experiments but checked for each Pareto-optimal path found at the terminal station whether it is bitonic or not.

Observations. It turns out that, on average, .84 percent of all Pareto-optimal paths are bitonic.

Experiment 6. *What is the number of Pareto-optima in data from other applications where our assumptions on the correlation between the cost functions do not hold?*

We used the digital elevation models from Austria and Scotland [ ] [2] where the underlying graphs are bidirected grid graphs, and one optimizes the accu-

---

[2] The data is available from `http://www.mpi-sb.mpg.de/~mark/cnop` .

mulated height difference vs. the distance travelled. These length functions are only loosely correlated.

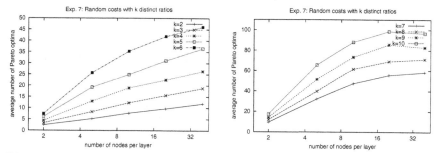

Fig. 5. The average number of Pareto optima for layer graphs with $k = 2, \ldots, 10$ different cost ratios

Observations. See Table 2. The maximal number of labels as well as the average number of labels per node is significantly larger than in the train graph instances (note that the graph sizes are by orders of magnitude smaller). Indeed, for the two larger instances, the memory requirements for the computation were in the range of 2GB.

4   Experiments on Artificial Instances

We have seen in Section 2 that it is possible to construct an artificial class of worst case examples with as many as $O(n^{k-1})$ different Pareto optima for graphs on n nodes with only k monotone increasing cost ratios.

Experiment 7. *What is the number of Pareto optima in graphs with k randomly chosen cost ratios and randomly chosen arc lengths?*

To study this question, we construct complete layered graphs with k edge layers and k monotonously increasing ratios $r[i]$ in the following way. The first ratio is normalized to $r[1] := 1$. For $1 < i \le k$, we chose an integral random number s_i uniformly distributed in the interval [1,10], and set $r[i] := r[i-1] + s_i$. For each arc a, we set the cost components to $(\ell(a), r\ell(a))$, where r denotes the cost ratio of the layer a belongs to and $\ell(a)$ is a randomly chosen length from the uniform distribution in the interval [1,100]. For fixed number of nodes n and fixed number of cost ratios k, we take 100 different series of cost ratios and combined them with 200 different sets of edge lengths, giving 20,000 different instances for each combination of n and k.

Observations. In Figure 5, each data point gives the number of Pareto-optimal labels at the terminal where we have taken the average over the 20,000 instances of the same size. For all k, the increase of the number of Pareto optima at the terminal is clearly sublinear in the number of nodes per layer. This is in sharp contrast to our artificially constructed worst case examples with as many as n^{k-1} different Pareto optima.

Acknowledgments

The German train schedule data and the queries have been made available to us by the TLC Transport-, Informatik- und Logistik-Consulting/EVA-Fahrplanzentrum, a subsidiary of the Deutsche Bahn AG. Both authors wish to thank Frank Schulz for his help in preparing the train graph data.

The first author was partially supported by grant We1943/3-1 of the Deutsche Forschungsgemeinschaft (DFG).

References

1. J. Brumbaugh-Smith and D. Shier. An empirical investigation of some bicriterion shortest path algorithms. *European Journal of Operations Research*, 43:216–224, 1989. 187
2. M. Ehrgott and X. Gandibleux. An annotated biliography of multiobjective combinatorial optimization. *OR Spektrum*, pages 425–460, 2000. 186
3. P. Hansen. Bicriteria path problems. In G. Fandel and T. Gal, editors, *Multiple Criteria Decision Making Theory and Applications*, volume 177 of *Lecture Notes in Economics and Mathematical Systems*, pages 109–127. Springer Verlag, Berlin, 1979. 186
4. O. Jahn, R. H. Möhring, and A. S. Schulz. Optimal routing of traffic flows with length restrictions. In K. Inderfurth et al., editor, *Operations Research Proceedings 1999*, pages 437–442. Springer, 2000. 186
5. K. Mehlhorn and G. Schäfer. A heuristic for Dijkstra's algorithm with many targets and its use in weighted matching algorithms. In *Proceedings of 9th Annual European Symposium on Algorithms (ESA'2001), to appear*. 2001. 192
6. K. Mehlhorn and M. Ziegelmann. Resource constrained shortest paths. In *Proceedings of 8th Annual European Symposium on Algorithms (ESA'2000)*, volume 1879 of *Lecture Notes in Computer Science*, pages 326–337. Springer, 2000. 186
7. K. Mehlhorn and M. Ziegelmann.   CNOP — a package for constrained network optimization. In *3rd Workshop on Algorithm Engineering and Experiments (ALENEX'01)*. 2001. 186, 194
8. J. Mote, I. Murthy, and D. L. Olson. A parametric approach to solving bicriterion shortest path problems. *European Journal of Operations Research*, 53:81–92, 1991. 186, 187
9. F. Schulz, D. Wagner, and K. Weihe. Dijkstra's algorithm on-line: An empirical case study from public railroad transport. In *Proceedings of 3rd Workshop on Algorithm Engineering (WAE'99)*, volume 1668 of *Lecture Notes in Computer Science*, pages 110–123. Springer, 1999. 188
10. A. J. V. Skriver and K. A. Andersen. A label correcting approach for solving bicriterion shortest path problems. *Computers and Operations Research*, 27:507–524, 2000. 186
11. D. Theune. *Robuste und effiziente Methoden zur Lösung von Wegproblemen*. Teubner Verlag, Stuttgart, 1995. 186

12. A. Warburton. Approximation of pareto optima in multiple-objective shortest path problems. *Operations Research*, 35:70–79, 1987. 186

Author Index

Lecture Notes in Computer Science

For information about Vols. 1–2048
please contact your bookseller or Springer-Verlag